# Crystal Engineering

## How molecules build solids

# Crystal Engineering

How molecules build solids

**Jeffrey H Williams**

Morgan & Claypool Publishers

Rights & Permissions
To obtain permission to re-use copyrighted material from Morgan & Claypool Publishers, please contact info@morganclaypool.com.

ISBN    978-1-6817-4625-8 (ebook)
ISBN    978-1-6817-4626-5 (print)
ISBN    978-1-6817-4627-2 (mobi)

DOI    10.1088/978-1-6817-4625-8

Version: 20170901

IOP Concise Physics
ISSN 2053-2571 (online)
ISSN 2054-7307 (print)

A Morgan & Claypool publication as part of IOP Concise Physics
Published by Morgan & Claypool Publishers, 1210 Fifth Avenue, Suite 250, San Rafael, CA, 94901, USA

IOP Publishing, Temple Circus, Temple Way, Bristol BS1 6HG, UK

*As always, for Bruce Taylor Currie*

# Contents

# Preface

There are more than 20 million chemicals in the literature, with new materials being synthesized each week. Most of these molecules are stable, and the 3-dimensional arrangement of the atoms in the molecules, in the various solids may be determined by routine x-ray crystallography. When this is done, it is found that this vast range of molecules, with varying sizes and shapes can be accommodated by only a handful of solid structures.

This limited number of architectures for the packing of molecules of all shapes and sizes, to maximize attractive intermolecular forces and minimize repulsive intermolecular forces, allows us to develop simple models of what holds the molecules together in the solid. In this volume we look at the origin of the molecular architecture of crystals; a topic that is becoming increasingly important and is often termed, crystal engineering. Such studies are a means of predicting crystal structures, and of designing crystals with particular properties by manipulating the structure and interaction of large molecules. That is, creating new crystal architectures with desired physical characteristics in which the molecules pack together in particular architectures; a subject of particular interest to the pharmaceutical industry.

# Introduction: Crystal engineering

Chemists are busy people; there are over 20 million chemicals known in the literature, with many new molecules being synthesized each and every week. Most of these molecules are stable enough to be studied as they change state from solid to liquid, or solid to gas as the temperature is varied. And this huge number of molecules come in an appropriately wide variety of shapes and sizes. However, when it comes to how this huge diversity of molecules are packed together in solids, under the influence of the attractive electromagnetic forces that exist between all molecules, it is found that there are only a limited number of ways of ordering the stable packing of these variously sized and shaped molecules. This limited number of architectures for the packing of molecules to maximize attractive intermolecular forces and, consequently, minimize repulsive intermolecular forces, is the subject of this volume. We will look at the molecular architecture of crystals.

Looking at intermolecular interactions via a consideration of the detailed 3-dimensional arrangements of the molecules that constitute the solid, is not the usual route for studying intermolecular forces. However, the stable structure of a solid is the result of all possible interactions between the molecules that constitute that solid; that is, binary-interactions plus tertiary-interactions plus... N-body interactions, and I hope that my choice of the molecules reveals something of the contribution of these various terms that make up the intermolecular potential to the observed solid structure.

As the Nobel laureate Francis Crick said, 'If you want to understand function, study structure', and we will therefore consider the structure or internal architecture of a number of materials as determined by x-ray and neutron crystallography. Some of the 27 materials that I discuss in this volume have interesting and unusual bulk properties; for example, high-temperature superconductivity, birefringence, or laser activity. The data required to construct these 3-dimensional structures is stored in databases such as the Cambridge Structural Database (www.ccdc.cam.ac.uk), which presently contains the structure of over 875 000 covalently-bonded crystals; with new structures being added continually. These structures represent data accumulated by studies on single-crystals and on powders over the period 1914 (when the first crystal structure was published) to the present day.

In our study, we will concentrate on the weak, non-bonding intermolecular forces that lead gases to condense into liquids, and liquids to transform into ordered solids as the temperature is lowered, or the pressure increased. It is these weak forces that cause covalently bonded organic compounds to crystallize in the manner they do, and it is the study and comparison of these crystal structures that allow us to comprehend the nature of the underlying intermolecular forces.

In one of the earlies conjectures about the architecture of solids; the Roman poet and philosopher Titus Lucretius Carus (c 99–c 55 BCE) said in his *De Rerum Natura*, 'Those things, whose textures fall so aptly contrary to one another that hollows fit solids, each in the one and the other, make the best joining'. This qualitative observation of how macroscopic objects, or jigsaw puzzles fit together has its modern counterpart in the observations of the Russian physicist Alexander I Kitaigorodsky who wrote, 'The projections of one molecule get into the "hollows" of adjacent molecules, so that the molecules are closely packed with the minimum [number of] voids between them'; that is, as the crystallographer Jack Dunitz has said, 'As far as the packing energy is concerned, empty space is wasted space'. Thus, we observe that molecules pack so as to avoid empty space; but why? What forces the molecules together to minimize intervening vacuum? This is what we will explore in this volume.

At the beginning of the last century, hardly anything was known about the arrangement of molecules in crystals. Indeed, the external point group symmetries of crystals had been determined from the examination of the symmetry of their external faces. As far as the internal structure of the crystals was concerned, the concept of space lattices had been introduced in the 18th century to explain the constancy of interfacial angles in differently shaped crystals of the same compound. The geometric theory of space lattices was complete by the end of the 19th century, culminating in the almost simultaneous recognition that there are only a finite number (230) of ways of combining elements of point symmetry with translational symmetry to form space groups. The mathematical theory of crystals may have been complete, but nothing was known about the underlying structure of the repeating units within those complex shapes.

This all changed with the discovery of x-ray diffraction by crystals. The first x-ray studies were of simple ionic crystals of high symmetry, such as the alkali halides (cubic), but it was not long before complex minerals, such as mica, $KAl_3Si_3O_{10}(OH)_2$ (monoclinic) were being studied. With metals, the x-ray diffraction studies of the American physicist Albert Hull showed that the arrangement of atomic nuclei in many metallic crystals corresponds to the close-packed structures of spheres.

But for all the technical advances, questions about how organic molecules attracted or repelled one another took longer to be asked, and still longer to be answered. Such quantitative relationships as we have to explain the condensed phases came from experimental observations and ideas about the shape and size of molecules, although the nature of the molecular entities and of the forces acting among them were only approximately known. Today we still talk about van der Waals molecular volumes and van der Waals radii, and also about van der Waals forces without defining too closely what they mean.

In contrast to inorganic compounds, even simple organic compounds typically form crystals of low symmetry. Thus, crystals of anthracene examined by x-ray diffraction in 1920, were found to be monoclinic and thus intractable by the methods then in use. According to folklore, the idea for the ring structure of benzene came to German chemist Friedrich August Kekulé (1829–1896) in a daydream of 1865, in which he envisioned a snake eating its own tail. Kekulé's daydream led him to propose that the structure contained a six-membered ring of carbon atoms with alternating single and double bonds. But questions still remained; for example, in what configuration was this ring? Was it puckered, bowed, or flat? Did the molecule have three distinct double bonds? Most chemists subscribed to the theory that benzene was flat, but it was not until British crystallographer Kathleen Lonsdale

(1903–1971) began her research in 1929 that the mystery was finally solved. Unlike benzene, hexamethylbenzene has just one molecule per unit cell, making it easier to distinguish the orientation of the molecule's central benzene ring. Lonsdale's x-ray crystallography experiments unequivocally demonstrated that the benzene ring was not only flat, but also had an evenly distributed cloud of electrons; there were no single or double carbon–carbon bonds. Because the benzene ring is the foundation of aromatic compounds, Lonsdale's discovery made it possible to advance the chemistry of aromatic molecules, and laid the basis for the modern form of organic chemistry, and for molecule and crystal design. As Lonsdale's head of department, Christopher K Ingold commented on her discovery, 'The calculations must have been dreadful... but one structure like this brings more certainty into organic chemistry than generations of activity by us [physicist] professionals'.

In quantum mechanics, there is no such thing as empty space. As far as the physics of intermolecular interactions is concerned, what matters is the nature and strength of the fields of force generated by the electrons and nuclei. In inorganic structures, the strong Coulombic fields exerted by the cations and anions were comparatively easy to understand, and atomic cohesion in such crystals seemed to present no fundamental problems. Similarly, as an obvious extension, it was soon realized that the interaction between molecules with permanent dipole moments, such as water, is subject to analogous Coulombic fields and may be attractive or repulsive, depending on the mutual orientation of the molecules. The nature of the cohesive forces among neutral non-polar molecules remained elusive for a longer time. For example, no theory based on classical mechanics and electrostatics could possibly reproduce the lattice energy of methane or the inert gases. This may be a small effect, but it is undoubtedly present, as witnessed by the condensation of these gases and the solidification of the liquid at sufficiently low temperature. The mysterious missing term, the dispersion energy, had to await the advent of quantum mechanics and Fritz London who first described dispersion interactions. Thus, explaining weak intermolecular interactions and pointing out that for these short-range forces to be effective, molecules must be in close contact; bumps into hollows, with as little empty space as possible. It was Kitaigorodsky's achievement, in the early-1960s, to put these concepts on a systematic footing beginning with a critical survey of organic crystal structures.

X-ray crystallography is not a difficult or obscure branch of science; on the contrary it is now a routine procedure, and measurements are today automated and the data analysis has become a *black-box* technology. This volume seeks to show that the beautiful and fascinating, detailed 3-dimensional pictures of closely-packed molecules tells us a great deal about molecular interactions, and the bulk behavior of solids. However, before we start exploring crystal architecture we will consider some of the basics of chemical structure, bonding and the nature of intermolecular forces. First, we will look at how atoms bond to form molecules and how electrons are distributed in those molecules. The reader will then discover that considerations of symmetry play a central role in classifying the structure of solids as well as in rationalizing the properties of crystalline materials.

Attentive readers will have noticed that it is barely a century since the very first crystal structure determination using x-ray diffraction, 1914. Not surprisingly, this anniversary generated a number of review articles. One of the most interesting is to be found at http://cen.xraycrystals.org/introduction.html, and comprises a set of highlights of crystallography over the last century as chosen by *Chemical&Engineering News*, the news magazine of the American Chemical Society.

# Acknowledgement

With this volume, *Crystal Engineering: How molecules build solids*, I should like to thank Dr Jeremy Karl Cockcroft, Chemistry Department, University College London, for all his help. I have known Jeremy Cockcroft since we both worked as experimental scientists at the Institute Laue-Langevin, Grenoble; where I was a solid and liquid-state physicist and Jeremy was a neutron crystallographer. Over the years, Jeremy has taught me a great deal about crystallography and about how a crystallographer looks at molecular interactions; and his help has been central in writing this volume. In addition, he has also got me back in the laboratory doing research. Earlier this year, I published my first technical paper in the peer-reviewed literature since 1993... back to the future.

# Author biography

**Jeffrey Huw Williams**

Jeffrey Huw Williams was born in Swansea, Wales, on 13 April 1956, and gained a PhD in chemical physics from Cambridge University in 1981.

Subsequently, his career as a research scientist was in the physical sciences. First, as a research scientist in the universities of Cambridge, Oxford, Harvard and Illinois, and subsequently as an experimental physicist at the Institute Laue-Langevin, Grenoble, which still remains one of the world's leading centres for research involving neutrons; particularly, neutron scattering. During this research career, he published more than sixty technical papers and invited review articles in the peer-reviewed literature. However, after much thought he left research in 1992 and moved to the world of science publishing and the communication of science by becoming the European editor for the physical sciences for the AAAS's *Science*.

Subsequently, he was Assistant Executive Secretary of the International Union of Pure and Applied Chemistry; the agency responsible for the world-wide advancement of chemistry through international collaboration. Most recently, 2003–2008, he was the head of publications at the *Bureau international des poids et mesures* (BIPM), Sèvres. The BIPM is charged by the Metre Convention of 1875 with ensuring world-wide uniformity of measurements, and their traceability to the International System of Units (SI). It was during these years at the BIPM that he became interested in, and familiar with the origin of the Metric System, its subsequent evolution into the SI, and the coming transformation into the Quantum-SI.

Since retiring, he has devoted himself to writing; in 2014, he published *Defining and Measuring Nature: The make of all things* in the IOP Concise Physics series. This publication outlined the coming changes to the definitions of several of the base units of the SI, and the evolution of the SI into the Quantum-SI. In 2015, he published *Order from Force: A natural history of the vacuum* in the IOP Concise Physics series. This title looks at intermolecular forces, but also explores how ordered structures, whether they are galaxies or crystalline solids, arise via the application of a force. Then in 2016, he published *Quantifying Measurement: The tyranny of number*, again the IOP Concise Physics series. This title is intended to explain the concepts essential in an understanding of the origins of measurement uncertainty. No matter how well an experiment is done, there is always an uncertainty associated with the final result—something that is often forgotten.

**IOP** Concise Physics

# Crystal Engineering
How molecules build solids
**Jeffrey H Williams**

# Chapter 1

## Holding things together

The chemical bond is the heart of chemistry. If we can understand what holds atoms together in molecules, we may begin to understand why, under certain conditions, existing arrangements of atoms may change to become different arrangements; that is, to understand the basis of chemical transformation. And if we understand molecular structure, we may hope to identify the mechanism of chemical change. In addition, if the structure and bonding of molecules is understood, one may also attempt to model the non-bonding interactions that arise between molecules, and which are responsible for condensed phases.

All this is, of course, a tall order; so let us first simplify things. In quantum chemistry and molecular physics, the Born–Oppenheimer (BO) approximation is the assumption that the motion, or dynamics of the atomic nuclei and the electrons in a molecule can be separated. It is the means by which a meaningful approach can be made to using quantum mechanics to explain molecular phenomena. The approach is named after the German physicist and one of the originators of quantum mechanics Max Born (1882–1970) and one of his doctoral students, the American physicist and 'father of the atom bomb' J Robert Oppenheimer (1904–1967). In mathematical terms, the BO approximation allows the wavefunction of a molecule to be factorized into its electronic and nuclear (that is, vibrational, rotational and translational) components,

$$\Psi_{molecule} = \psi_{electronic} \times \psi_{nuclear}.$$

Computation of the energy and the wavefunction of an average-size molecule is simplified by this approximation; for example, the benzene molecule consists of 12 nuclei and 42 electrons. The time-independent Schrödinger equation, which must be solved to obtain the energy and wavefunction of this molecule ($\Psi_{molecule}$), is a partial differential eigenvalue equation in 162 ($3 \times 54$) variables; the spatial coordinates of the electrons and the nuclei. The BO approximation makes it possible to compute the wavefunction in two, less complicated consecutive steps. This approximation

doi:10.1088/978-1-6817-4625-8ch1

was proposed in 1927, in the glory days of quantum mechanics and is still indispensable in quantum chemistry. The success of the BO approximation is due to the significant difference between nuclear and electronic masses[1].

## 1.1 Covalent bonds

A covalent bond is a chemical bond that involves the sharing of electron pairs between atoms. Where each atom contributes one electron to the intervening covalent bond. For many molecules, the sharing of electrons to form covalent bonds allows each atom to attain the equivalent of a full outer shell of electrons; this being a stable electronic configuration. The term *covalent bond* dates from the 1930s, and the prefix co-signifies shared or partnered; thus a co-valent bond means that the atoms share valence electrons; that is, electrons in the outer-most atomic orbitals.

In the simplest of all stable molecules, $H_2$, the hydrogen atoms share the two electrons via a single covalent bond. However, covalent bonding does not necessarily require that the two atoms be of the same element, only that they have a comparable capacity for pulling electrons to themselves; that is, a comparable electronegativity (a heuristic atomic quantity devised by the Swedish chemist Jöns Jacob Berzelius (1779–1848), and put on a quantitative basis in 1932 by the American chemist Linus Pauling (1901–1994).

The idea of covalent bonding can be traced back to Gilbert N Lewis (1875–1946), who in 1916 first described the sharing of electron pairs between atoms. He introduced the Lewis notation in which valence electrons are represented as dots around the atomic symbols in a chemical formula. Pairs of electrons located between atoms representing covalent bonds, and multiple pairs representing multiple bonds, such as double bonds and triple bonds. Lewis proposed that an atom forms enough covalent bonds so as to form a full (or closed) outer electron shell. In the covalent bonding seen in methane ($CH_4$), for example, the carbon atom has a valence of four and is, therefore, surrounded by eight electrons; four from the carbon atom and four from the four hydrogen atoms to which it is bonded. Each hydrogen atom has a valence of one and is surrounded by two electrons; that is, its own $1s^1$ electron plus one electron from the carbon.

Atomic orbitals (except *s*-orbitals, which are spherically symmetric) have specific directional properties leading to different types of covalent bonds when they overlap with similar orbitals on other atoms. Covalent bonding includes many kinds of orbital interactions that lead to variations of a simple covalent bond. For example, atomic *s*-orbitals can form a covalent bond, as can atoms with single electron in a *p*-orbital. But if there are three electrons in an atom's *p*-orbitals, there is the possibility of $\sigma$-bonding and $\pi$-bonding in a bonded homonuclear pair; these interactions lead to single, double or even triple covalent bonds as in: ethane (a single $\sigma$ covalent C–C bond), ethylene (a $\sigma$ covalent C–C bond and a single $\pi$ covalent C–C bond) and acetylene (a $\sigma$ covalent C–C bond and two $\pi$ covalent C–C bonds).

---

[1]A factor, of order, 1833; the ratio of the mass of the proton to the mass of the electron.

Covalent bonds are also affected by the relative electronegativities of the bonded atoms, which determines the chemical polarity of the bond. Two atoms with equal electronegativity will form a non-polar covalent bond; for example, $H_2$, $N_2$, and $O_2$. An unequal electronegativity creates a polar covalent bond such as HF and CO. Such heteropolar bonds lead to the presence of electric dipole moments in the molecule; that is, there is a difference between the centre of the distributions of positive and of negative charge within the molecule. However, a permanent molecular polarity also requires geometric asymmetry, or else the dipoles may cancel out resulting in a non-polar molecule; as in benzene, where the vector sum of the six bond dipole moments is zero.

There are an enormous range of structures that may be generated using covalent bonds, including molecules that are stable in both the gaseous and condensed phases. Yet, even if individual molecules are made up of covalently bound atoms and even when they have an asymmetry of charge, there are only weak interactions between the individual molecules. For example, a typical bond energy (the energy required to break a covalent bond, or the energy liberated when such a bond forms) for a C–H covalent bond is large at 439 kJmol$^{-1}$ in methane, but a material like methane is a gas at room temperature as there are only weak interactions between the individual molecules; and benzene, which contains six strong C–C bonds (each with a strength of about 350 kJ mol$^{-1}$) and six strong C–H bonds is a volatile liquid at room temperature. Thus, molecules made up of covalent bonds retain their individuality in the condensed phase.

## 1.2 Ionic bonds

Ionic bonding involves the electrostatic attraction between oppositely charged ions, and is the primary interaction occurring in ionic compounds. The ions are atoms that have either gained one or more electrons (anions, which are negatively charged), or atoms that have lost one or more electrons (cations, which are positively charged). This transfer of electrons is known as electrovalence in contrast to covalence. In the simplest case, the cation is a metal atom and the anion is a non-metal atom, but ions can be of a more complex nature; for example, molecular ions like $NH_4^+$ or $SO_4^{2-}$. Again, we may think of an ionic bond as the transfer of electrons from one atom to another atom so as to allow both atoms to obtain a full valence shell.

Ionic compounds reveal their ionic character by being able to conduct electricity when molten, or in aqueous solution. And because of the very powerful electrostatic forces generated over molecular dimensions by a transferred electron, ionic compounds generally have a high melting point, depending upon the charge of the ions present. The higher the electric charge ($Na^+$, $Ca^{2+}$, $Al^{3+}$) present on the constituent ions, the stronger the cohesive forces that bind the ions together, and the higher the melting point. This also makes such materials soluble in water, as the electric dipole moments of the water molecules are strongly attracted to the electrically-charged ions.

Ionic compounds in the solid-state are able to form giant lattice structures, built up of repeated subunits. The two principal factors in determining the form of the lattice are the relative charges of the ions and their relative sizes. Some structures are adopted by a number of compounds; for example, the structure of rock-salt, or sodium chloride is also adopted by many other alkali metal halides and binary oxides such as MgO. Linus Pauling formulated heuristic rules, by examining a vast number of materials, to provide guidelines for predicting and rationalizing the crystal structures of ionic crystals.

Ions in lattices of ionic compounds are spherical; however, if the positive ion is small and/or highly charged, it will distort the electron cloud of neighbouring negative ions, an effect summarised in Fajans' rules of 1923 (named for the Polish chemist Kazimierz Fajans). This polarization of the negative ion leads to a build-up of extra charge density between the two nuclei; that is, to a partial covalency. Larger negative ions are more easily polarized, but the effect is usually important only when cations with charges of 3+ are involved. However, 2+ ions and even 1+ show some polarizing power when their sizes are small; for example, $BeF_2$ is mostly covalent, and LiI is more ionic but has a significant covalency.

## 1.3 Comparison of ionic and covalent bonding

In ionic bonding, electrically-charged atoms are bound by the electrostatic attraction of oppositely-charged ions, whereas, in covalent bonding, atoms are bound together by the sharing of electrons to attain stable configurations of electrons. In covalent bonding, the molecular geometry around each atom is determined by valence shell electron pair repulsion rules (VSEPR), whereas, in ionic materials, the geometry follows maximum packing rules (the most effective way of packing pairs of charged-particles into a solid framework). One could say that covalent bonding is more directional in the sense that the energy penalty for not adhering to the optimum bond angles is large, whereas ionic bonding has no such penalty. In ionic solids there are no shared electron pairs to repel each other; the ions should simply be packed as efficiently as possible to maximize electrostatic attraction and to minimize electro-static repulsion. This often leads to much higher coordination numbers; in NaCl, each ion has six ionic bonds and all the bond angles are 90 degrees; that is, octahedral coordination. In CsCl, because the $Cs^+$ ion is much bigger than $Na^+$, the coordination number is 8. By comparison, carbon forms a maximum of four bonds and they are disposed tetrahedrally.

Purely ionic bonding cannot exist, as the proximity of the entities involved in the bonding allows some degree of the sharing of the electron density between them. Therefore, all ionic bonding has some covalent character. The larger the difference in electronegativity between the two atoms involved in the bonding, the more ionic (polar) it is. In general, when ionic bonding occurs in the solid state, it is not possible to speak about a single ionic bond between two individual atoms, because the cohesive forces that keep the lattice together are of a more collective nature. This is quite different in the case of covalent bonding, where we can often speak of a distinct bond localized between two particular atoms, oriented in a particular direction.

## 1.4 Non-bonding interactions

Non-polar molecules (including closed-shell atoms, such as neon and argon) attract one another even though the interacting molecules do not possess permanent dipole moments, as demonstrated by the condensation of these materials. The interaction between non-polar molecules arises from the transient dipoles that all atoms and molecules possess as a consequence of fluctuations in the instantaneous positions of the electrons found in those molecules. To appreciate the origin of this purely quantum mechanical interaction, suppose that the electrons in one molecule instantly move into an arrangement that gives the molecule an instantaneous, or fluctuating dipole moment $\mu_1$. This dipole generates an electric field that polarizes a neighbouring molecule, and induces in that molecule an instantaneous dipole moment $\mu_2$. The two dipoles attract each other and the potential energy of the pair is lowered. This interaction is often termed the dispersion interaction, or the London interaction (for Fritz London).

Polar molecules also interact by a dispersion interaction. However, as such molecules possess a permanent dipole moment, it is this intermolecular interaction that will dominate over the instantaneous dispersion interaction. Consequently, the interaction between two water molecules will be dominated by the electrostatic interaction of the large permanent electric dipole moments, but the interaction of two argon atoms will consist only of the dispersion interaction (the boiling point of argon is 87 K, and that of water is 373 K).

## 1.5 Hydrogen bonding

The dispersion interaction is universal in that it is exhibited by all molecules. However, there is a type of interaction possessed by molecules that have a particular chemical constitution. A hydrogen bond is an attractive interaction between two species that arises from a link ($\cdots$) of the form A–H$\cdots$B, where A and B are strongly electronegative elements and B possesses a lone pair of electrons. Hydrogen bonding is conventionally regarded as being limited to N, O, and F but, if B is an anionic species (such as $Cl^-$), it may also participate in hydrogen bonding. A hydrogen atom attached to a carbon atom can also participate in hydrogen bonding when the carbon atom is bound to electronegative atoms, as is the case in chloroform, H–$CCl_3$ (this point will be considered later with regard to the interaction between C–F and C–H bonds; that is C–F$\cdots$H–C).

In a hydrogen bond, the electronegative atom not covalently attached to the hydrogen is termed the proton acceptor, whereas the atom covalently bound to the hydrogen is termed the proton donor. In the donor molecule, the electronegative atom attracts the electron density from around the hydrogen nucleus of the donor group, and leaves the atom with a positive partial-charge. Because of the small size of hydrogen relative to other atoms, the resulting charge, though only partial, represents a significant charge density. A hydrogen bond results when this strong positive charge-density attracts a lone pair of electrons on another atom, which then becomes the hydrogen-bond acceptor.

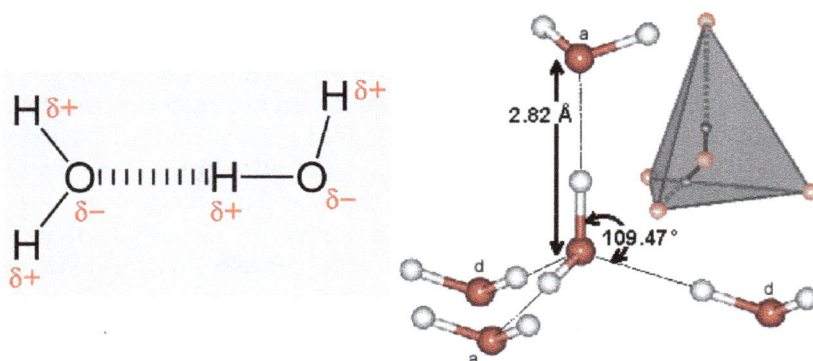

**Figure 1.1.** Some of the features of a 'classic' strong hydrogen bond as exemplified in liquid water and ice.

The hydrogen bond is sometimes described as an electrostatic dipole–dipole interaction; however, when we are speaking of strong hydrogen bonds it also has some features of covalent bonding; it is directional and can be strong, produces interatomic distances shorter than the sum of the van der Waals radii, and usually involves a limited number of interaction partners, which can be interpreted as a type of valence. These covalent features are more substantial when acceptors bind hydrogen atoms shared with more electronegative donors.

The partially covalent nature of a hydrogen bond raises a question as to which molecule or atom does the hydrogen nucleus belong, and which moiety should be labelled donor and which acceptor? Usually, this is clearly seen on the basis of interatomic distances in the X–H$\cdots$Y system, where the dots represent the hydrogen bond: the X–H distance is typically 1.1 Å, whereas the H$\cdots$Y distance is between 1.6 and 2.0 Å. Liquids that display strong hydrogen bonding, such as the archetype water, are termed associated liquids. Figure 1.1, give the classic features of hydrogen bonds, as seen in water.

Hydrogen bonds can vary greatly in strength; from weak (1–2 kJ mol$^{-1}$) to strong (161.5 kJ mol$^{-1}$ in the ion HF$_2^-$. Some typical values are: F–H$\cdots$F (161.5 kJ mol$^{-1}$), O–H$\cdots$N (29 kJ mol$^{-1}$), O–H$\cdots$O (21 kJ mol$^{-1}$), N–H$\cdots$N (13 kJ mol$^{-1}$) and N–H$\cdots$O (8 kJ mol$^{-1}$).

The length of hydrogen bonds depends on bond strength, temperature, and pressure. The bond strength itself is dependent on temperature, pressure, bond angle, and environment (usually characterized by local dielectric constant). The typical length of a hydrogen bond in water is 1.97 Å. The ideal bond angle depends on the nature of the hydrogen bond donor[2].

---

[2]In the gas-state, high-resolution microwave spectroscopy of molecular dimers gives a value of 2.14 Å between the oxygen atom of $CO_2$ and the hydrogen atom of HCl in the linear hydrogen-bonded dimer O–C–O$\cdots$H–Cl, and a distance of 3.17 Å between the plane of the benzene molecule and the fluorine of HF in the symmetric-top dimer benzene:HF where the hydrogen of the HF is pointing towards the centre-of-mass of the benzene ring.

## 1.6 Hybrid atomic orbitals and the shape of molecules

One of the more emblematic images in the Zodiac is the Centaur; a hybrid animal having the head, arms and torso of a man on the body and legs of a horse. The application of quantum mechanics to chemistry with a view to explaining the bonding and the shape of molecules also contains such hybrids. In quantum chemistry, one mixes atomic orbitals that have been hybridized in the same way that the Ancient Greeks mixed humans and animals to create hybrid mythological species. But whereas hybridizing and mixing orbitals to explain molecular structure is a much more recent myth than that of the Centaurs, it is still only an attempt to explain reality. There is no fully functioning quantum model of chemical bonding and of chemical structure; there are only approximate, hybrid methods.

The atomic orbitals needed to explain bonding in polyatomic systems need not necessarily be pure atomic orbitals; that is purely $s$-orbitals or purely $p$-orbits. Often, the bonding atomic orbitals have a character consisting of several types of atomic orbitals. Using quantum mechanical methods to combine the wavefunctions of atomic orbitals to construct new hybrid orbitals, more suitable to describe the bonding in molecules, is termed hybridization of atomic orbitals. For example, $sp$ hybrid atomic orbitals are possible states of an electron in an atom, especially when it is bonded to other atoms. These electron states have, for example, half the character of $2s$ and half of $2p$ orbitals, and there are two ways to combine the $2s$ and $2p$ atomic orbitals: $sp(1) = 2s + 2p$ or $sp(2) = 2s - 2p$.

These energy states, $sp(1)$ and $sp(2)$, each have a region of high electron probability, and the two atomic orbitals are located opposite to each other, centered on the atom. For example, the molecule H–Be–H is formed by the overlapping of two $1s$ orbitals from two H atoms and the two $sp$ hybridized orbitals of Be. The H–Be–H molecule is linear, as is the isolated (gas-phase) F–Be–F molecule. The unhybridized electronic configuration of Be is $1s^2 2s^2$, and one may think of the electronic configuration upon hybridization (before bonding) as $1s^2 sp^2$. The two electrons in the new $sp$ hybrid orbitals having the same energy as the unhybridized configuration.

In general, when two and only two atoms bond to a third atom and the third atom makes use of $sp$ hybridized orbitals, the three atoms form a linear molecule; for example, $sp$ hybrid orbitals are invoked to explain the shape of linear molecules such as F–Be–F, HCCH, HCN and O=C=O.

For $sp^2$ hybrid orbitals, the energy states of the valence electrons in atoms of the second period are in the $2s$ and $2p$ orbitals. If we mix two of the $2p$ orbitals with a $2s$ orbital, we generate three $sp^2$ hybrid orbitals. These three orbitals lie on a plane pointing to the vertices of an equilateral triangle; for example, BF$_3$, and the carbonate anion $CO_3^{2-}$. However, not all three $sp^2$ hybridized orbitals have to be used in bonding. One of the orbitals may be occupied by a pair of, or by a single electron. And these molecules are bent rather than linear; for example, SO$_2$.

Carbon is famously capable of generating four $sp^3$ hybrid orbitals and these are directed to the vertices of a regular tetrahedron, thus explaining a great deal of organic chemistry. Carbon atoms also makes use of the $sp^2$ hybrid orbitals in ethylene (also known as ethene), $H_2C=CH_2$. Here, the remaining $p$-orbital from each of the carbon overlap sideways to form an additional $\pi$-bond. As in the situation of $sp^2$ hybrid orbitals, one or two of the $sp^3$ hybrid orbitals may be occupied by non-bonding electrons; as in water and ammonia. The C, N and O atoms in $CH_4$, $NH_3$, $H_2O$ molecules use the $sp^3$ hybrid orbitals, however, a lone pair occupies one of the orbitals in $NH_3$, and two lone pairs occupy two of the $sp^3$ hybrid orbitals in $H_2O$. And there is electrostatic repulsion between these bonding and lone pairs of electrons, which is invoked to explain the shape of the molecules.

For $sp^3d$ hybrid orbitals, five hybrid orbitals formed when one $3d$, one $3s$, and three $3p$ atomic orbitals are mixed. When an atom makes use of five $dsp^3$ hybrid orbitals to bond to five other atoms, the geometry of the molecule is often a trigonal bipyramid; for example, the molecule $PClF_4$. In this molecule, the Cl atom takes up an axial position of the trigonal bipyramid. There are also structures in which the Cl atom may take up the equatorial position. The change in arrangement is accomplished by simply changing the bond angles. Again, some of the $dsp^3$ hybrid orbitals may be occupied by electron pairs.

The six $d^2sp^3$ hybrid orbitals result when two $3d$, one $3s$, and three $3p$ atomic orbitals are mixed. When an atom makes use of six $d^2sp^3$ hybrid orbitals to bond to six other atoms, the molecule takes the shape of an octahedron, as in $SF_6$. There are also cases that some of the $d^2sp^3$ hybrid orbitals are occupied by lone pair electrons.

## Further reading

For a general consideration of chemical bonding, there is no better source than the classic text by Peter W Atkins and Julio de Paula, *Physical Chemistry* (9th edition, 2010) Oxford University Press. For those readers seeking a more theoretical approach to the subject; particularly concerning the observed shapes of molecules, I recommend Peter W Atkins and Ronald Friedman, *Molecular Quantum Mechanics* (4th edition, 2005) Oxford University Press.

Crystal Engineering
How molecules build solids
**Jeffrey H Williams**

# Chapter 2

## Intermolecular electrostatics

There is a wealth of detail in the various forms of non-bonding molecular interactions displayed by solids formed of molecules containing covalently bound atoms. Some of these interactions are quite strong (relative to a hydrogen bond) and some of them are weaker. Molecules may be characterized by their permanent electric moments; that is, the distribution of charge in the molecule, which generate an electric field that distorts the electronic structures of neighbouring molecules.

In the study of molecular interactions, we are concerned with the question: what is the difference between the energy of a group of molecules and the energy of separate molecules, for fixed molecular positions and orientations? In solids, we are concerned with the interaction energies of simple molecules at distances from one another that are large compared to molecular dimensions. The aim is to describe the dependence of the interaction energy on the separations and orientations of the interacting molecules.

The Hamiltonian describing a molecule in weak interaction with a fixed set of external charges is:

$$H = H_0 - \mu_\alpha E_\alpha - (1/3)\Theta_{\alpha\beta}E_{\alpha\beta} - \cdots \tag{2.1}$$

where $H_0$ is the Hamiltonian for the free molecule, and $\mu_\alpha = \sum e_i r_i$ and $\Theta_{\alpha\beta} = \Theta_{\beta\alpha} = \frac{1}{2}\sum e_i(3r_{i\alpha}r_{i\beta} - r_i^2\delta_{\alpha\beta})$ are the dipole and quadrupole moment operators, respectively; $e_i$ is the $i$th element of charge at the point $r_i$ relative to an origin fixed at some point in the molecule[1]. $E_\alpha$ and $E_{\alpha\beta}$ are the electric field and the field-gradient at the origin due to the external charges.

---

[1] The Greek subscripts $\alpha$, $\beta$ denote vector or tensor components and can be equal to $x$, $y$, $z$; a repeated Greek subscript denotes a summation over all three Cartesian components, so that $(1/3)\Theta_{\alpha\beta}E_{\alpha\beta}$ is a sum over $\alpha$ and $\beta$ equalling $x$, $y$, and $z$ and so is a scalar quantity (here it represents the quadrupolar interaction with the field-gradient).

If the molecule is in the quantum state $\psi$, its energy $(W)$ for a fixed position and orientation is:

$$W = <\psi|H|\psi>$$
$$= W^{(0)} - \mu_\alpha^{(0)}E_\alpha - \frac{1}{2}\alpha_{\alpha\beta}E_\alpha E_\beta - (1/6)\beta_{\alpha\beta\gamma}E_\alpha E_\beta E_\gamma - \cdots \quad (2.2)$$
$$- (1/3)\Theta_{\alpha\beta}^{(0)}E_{\alpha\beta} - (1/3)A_{\alpha,\beta\gamma}E_\alpha E_{\beta\gamma} - \cdots.$$

where $\mu_\alpha^{(0)} = <\psi^{(0)}|\mu_\alpha|\psi^{(0)}>$ and $\Theta_{\alpha\beta}^{(0)} = <\psi^{(0)}|\Theta_{\alpha\beta}|\psi^{(0)}>$ are the permanent dipole and the permanent quadrupole moments of the molecules under consideration. Equation (2.2) represents the interaction of the various polarizabilities and hyperpolarizabilities of the molecules with the electric fields and electric field-gradients to which the molecules are subject. There are many more terms in equation (2.2) than given above; a full treatment of this problem is given in [1]. The total dipole and quadrupole moments of the molecule in the state $\psi$ are the expectation values of the operators, and are:

$$\mu_\alpha = -\frac{\partial W}{\partial E\alpha} = \mu_\alpha^{(0)} + \alpha_{\alpha\beta}E_\beta + \frac{1}{2}\beta_{\alpha\beta\gamma}E_\beta E_\gamma + \cdots (1/3)A_{\alpha,\beta\gamma}E_{\beta\gamma} + \cdots \quad (2.3)$$

$$\Theta_{\alpha\beta} = -3\frac{\partial W}{\partial E\alpha\beta} = \Theta_{\alpha\beta}^{(0)} + A_{\gamma,\alpha\beta}E_\gamma + \cdots \quad (2.4)$$

The second-rank tensor $\alpha$ is the familiar static polarizability, and $\beta$ is a hyperpolarizability describing deviations from a linear polarization law. If the molecular origin is the centre-of-symmetry, as it would be if it were the carbon atom of carbon dioxide then $\mu$, $\beta$ and $A$ are zero. And only the first, non-vanishing moment is independent of the choice of origin within the molecule. Thus, for an ion like $OH^-$, the dipole and quadrupole moments vary in the molecule depending upon the selected origin, and $\mu = 0$ at the centre of charge. Similarly, in HF and CO, the quadrupole moment depends on the selected origin, and is zero at the centre of dipole. In an uncharged, polar molecule, the dipole moment $\mu$ is independent of the origin.

The number of independent constants needed to describe the interaction energy (via equation (2.2)) with the external field is determined by the symmetry of the molecule. The number of constants required to specify the tensors $\mu$, $\Theta$, $\alpha$, $\beta$ and $A$ for a few molecular symmetry types is given in table 2.1. It can be shown that the maximum number of constants specifying a multipole moment of order $n$ is $(2n + 1)$; thus, one charge, three dipoles, five quadrupoles, seven octupoles are needed for the general molecule. The three dipoles could be given as $\mu_1$, $\mu_2$, and $\mu_3$ with respect to mutually perpendicular axes fixed in the molecule, or as a total dipole $\mu$ and two numbers specifying the orientation of this dipole with respect to the bonds of the molecule. The number of quadrupoles is one fewer than the number of polarizabilities $\alpha$, because the trace $\Theta_{\alpha\alpha} = 0$. We also see from table 2.1 that the polarizability is a universal property of matter, all atoms and molecules are polarizable, and hence capable of condensation.

**Table 2.1.** Number of constants needed to define the electrical properties of molecules of differing symmetry (see [2] for full details).

| Group Symbol | $\mu_\alpha$ | $\Theta_{\alpha\beta}$ | $\alpha_{\alpha\beta}$ | $\beta_{\alpha\beta\gamma}$ | $A_{\alpha,\beta\gamma}$ |
|---|---|---|---|---|---|
| $C_1$ | 3 | 5 | 6 | 10 | 15 |
| $C_{2v}$ (e.g. $H_2O$) | 1 | 2 | 3 | 3 | 4 |
| $D_{2h}$ (e.g. ethylene) | 0 | 2 | 3 | 0 | 0 |
| $C_{3v}$ (e.g. $NH_3$) | 1 | 1 | 2 | 3 | 3 |
| $D_{6h}$ (e.g. benzene) | 0 | 1 | 2 | 0 | 0 |
| $T_d$ (e.g. $CH_4$) | 0 | 0 | 1 | 1 | 1 |
| $O_h$ (e.g. $SF_6$) | 0 | 0 | 1 | 0 | 0 |
| $C_{\infty v}$ (e.g. CO) | 1 | 1 | 2 | 2 | 2 |
| $D_{\infty h}$ (e.g. $CO_2$) | 0 | 1 | 2 | 0 | 0 |
| Sphere (e.g. argon) | 0 | 0 | 1 | 0 | 0 |

When considering the interaction of a pair of molecules using models based on localized electrical moments interacting on the polarizabilities of neighbouring molecules, the molecules must be far apart; sufficiently far apart that electron exchange can be neglected.

## 2.1 Two interacting molecules

Let us look briefly at the electromagnetic forces that operate between molecules, and which give rise to the condensed phases. The ability to predict the solid-state packing of molecules, and to comprehend the observed molecular dynamics from knowledge of the electrical properties of the isolated molecules, is a goal that is much sought after, but is not easily realized. Although the strength of the various intermolecular interactions may be approximated, the problem is not straightforward. The utility of such modelling is the argument of the present work, and we will now look briefly at some structure–property relationships in solid benzene and in the binary adduct benzene:hexafluorobenzene. This latter, is the simplest member of a large class of layered organic compounds, variously referred to as binary adducts, charge-transfer complexes, or coplanar crystals.

The charge distribution of the simple aromatic molecules benzene and hexa-fluorobenzene are of some interest in modelling the architecture of solids. The symmetry of these two molecules tells us that all odd electrical moments (electric dipole moment and octupole moment) vanish and the first non-vanishing electrical moment is the molecular quadrupole moment, $\Theta$. In addition, the molecular symmetry allows only one independent component of $\Theta$, $\Theta_{zz} = -(\Theta_{xx} + \Theta_{yy})$, off-diagonal elements of $\Theta$ are zero. This quadrupole moments of benzene and hexafluorobenzene have been measured. Benzene ($C_6H_6$) has a large negative quadrupole moment ($-29 \pm 1.7 \times 10^{-40}$ C m$^2$) and hexafluorobenzene ($C_6F_6$) has a large positive quadrupole moment ($31.7 \pm 1.7 \times 10^{-40}$ C m$^2$) [3]. Simple mixing of equimolar quantities of these two fluids at room temperature, produces a solid which melts at 25 °C; that is, at a higher temperature than the two pure materials (both benzene and hexafluorobenzene melt at about 5.5 °C). This molecular adduct arises because of the strong electrostatic interactions between the benzene and hexafluorobenzene molecules that constitute the solid in equal quantities; these are the same forces that are responsible for the structure of the pure solids.

Consider the case of a pair of molecules, which have high symmetry (possessing a rotational axis, $C_n$, where $n \geqslant 3$) and are placed in the configuration defined in figure 2.1; we may then write for the attractive electrostatic interaction energy, $U_{el}$, as a function of intermolecular distance and orientation [1]:

$$
\begin{aligned}
U_{el}(R, \alpha, \varphi) = {} & \mu_1\mu_2 R^{-3}(2\cos\alpha_1\cos\alpha_2 + \sin\alpha_1\sin\alpha_2\cos\varphi) \\
& + (3/2)\mu_1\Theta_2 R^{-4}[\cos\alpha_1(3\cos^2\alpha_2 - 1) + 2\sin\alpha_1 \\
& \quad \sin\alpha_2\cos\alpha_2\cos\varphi] \\
& + (3/2)\mu_2\Theta_1 R^{-4}[\cos\alpha_2(3\cos^2\alpha_1 - 1) + 2\sin\alpha_1 \\
& \quad \cos\alpha_1\sin\alpha_2\cos\varphi] \\
& + (3/4)\Theta_1\Theta_2 R^{-5}(1 - 5\cos^2\alpha_1 - 5\cos^2\alpha_2 + 17\cos^2\alpha_1 \\
& \quad \cos^2\alpha_2 + 2\sin^2\alpha_1\sin^2\alpha_2\cos^2\varphi \\
& +16\sin\alpha_1\cos\alpha_1\sin\alpha_2\cos\alpha_2\cos\varphi)
\end{aligned}
\tag{2.5}
$$

where $\mu_1$ and $\mu_2$ are the permanent dipole moments of molecule one and molecule two, respectively, and $\Theta_1$ and $\Theta_2$ are the permanent quadrupole moments of molecules one and two, respectively. If, however, we are considering an interacting pair of non-polar molecules such as two benzene, or two hexafluorobenzene molecules, then $\mu_1 = \mu_2 = 0$, and we are left with the term varying as $R^{-5}$.

A consideration of the possible arrangements of the two interacting quadrupole moments (that is, an analysis of (2.5), when $\mu_1 = \mu_2 = 0$) tells us that the two orientations with the largest interaction energy are a slipped parallel and T-shaped arrangements of the molecules; that is, for benzene; as seen in figure 2.2

In the upper image of figure 2.2, the difference in the interaction energy of the two possible arrangements of quadrupolar molecules such as benzene, hexafluorobenzene, carbon dioxide, acetylene, or nitrogen, is small and is proportional to the angle between the main rotational axes of the two interacting molecules in the plane

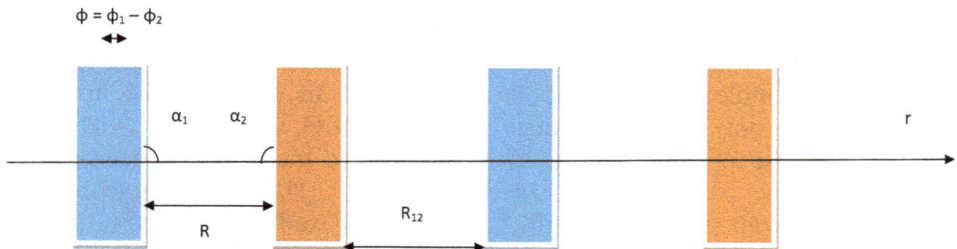

**Figure 2.1.** Schematic representation of the interaction of a series of molecules.

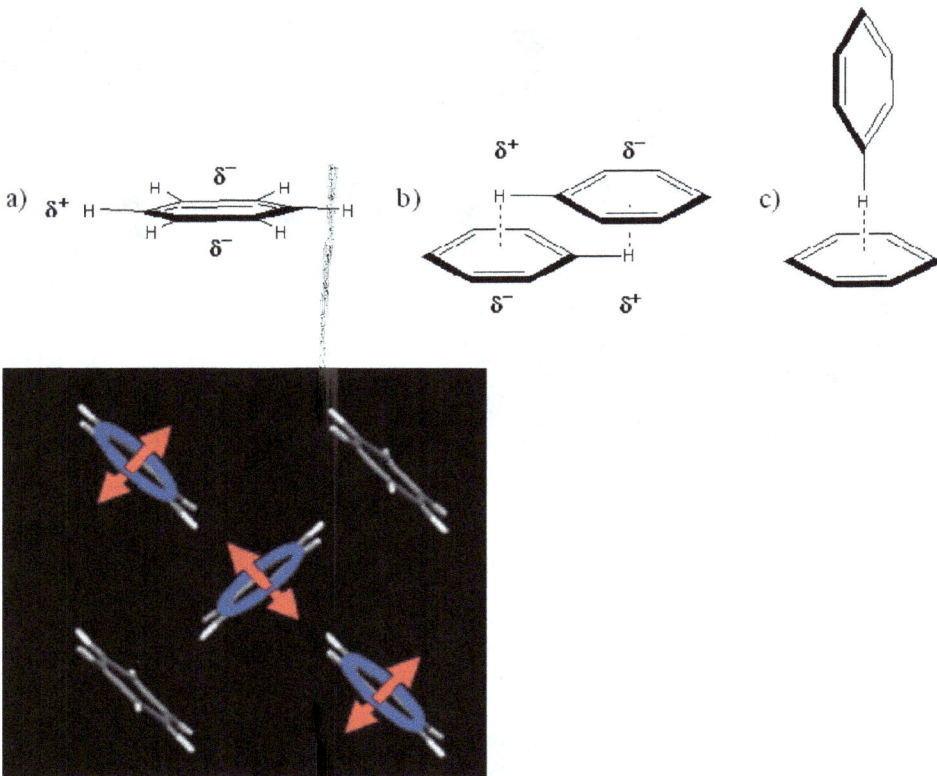

**Figure 2.2.** Interacting molecular quadrupole moments. Upper images: (a) the electronic structure of benzene represented as partial charges, (b) the slipped parallel attractive orientation of two benzene molecules ($\alpha_1 = 0$, $\alpha_2 = 0$ in equation (2.5)) and (c) the T-shaped attractive orientation of two benzene molecules ($\alpha_1 = 0$, $\alpha_2 = \pi$ in equation (2.5)). These relative positions are dictated by the interaction of the partial charges, and demonstrates why these two orientations are so favored in crystal architectures. (This image has been obtained by the author from the Wikimedia website https://en.wikibooks.org/wiki/Molecular_Simulation/Quadrupole-Quadrupole_ Interactions, where it is stated to have been released into the public domain. It is included within this article on that basis.) The lower coloured image gives a picture of how a few quadrupole moments of the same phase (the red arrow indicating the negative charge density above and below the plane (blue) of the molecule, which is the positively charged carbon ring) would align by intermolecular electrostatics. One can clearly see that the two preferred orientations of the benzene rings are electrostatic in origin.

containing the two molecules, see figure 2.1. As it happens, these are the two orientations of the benzene rings seen in crystalline benzene, see figure 2.3; that is, the architecture of solid benzene comprises benzene molecules in only these two orientations.

We can also see that when we take an equal number of positive and negative quadrupole moments, a new structure is generated, which is very different from the structure that results from the interaction of only positive, or negative quadrupole moments. In figure 2.4, we see the structure of the lowest temperature phase (IV) of the binary adduct benzene:hexafluorobenzene. Here, there is no 'herringbone'

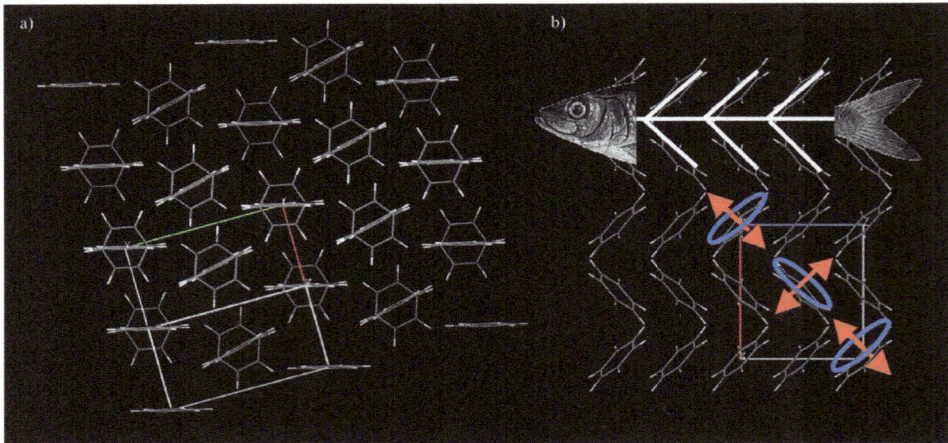

**Figure 2.3.** The structure of solid benzene: (a) the perpendicular orientation (T-shaped) of pairs of aromatic molecules and (b) the characteristic 'herringbone' structure of such quadrupolar solids (illustrated by the solid white lines in a fishbone arrangement). The negative electric quadrupole moments of benzene are represented in part (b) by a red arrow representing the concentration of negative charge along the normal to the plane of carbon atoms, with a blue circle corresponding to the depletion of negative charge. The x-ray structural data used to construct this image was taken from the Cambridge Structural Database, and are the work of Cox *et al* [4]. (Image with thanks from https://www.researchgate.net/figure/46271611_fig1_Figure-3-Two-distinct-snapshots-showing-the-crystalline-structure-of-solid-benzene).

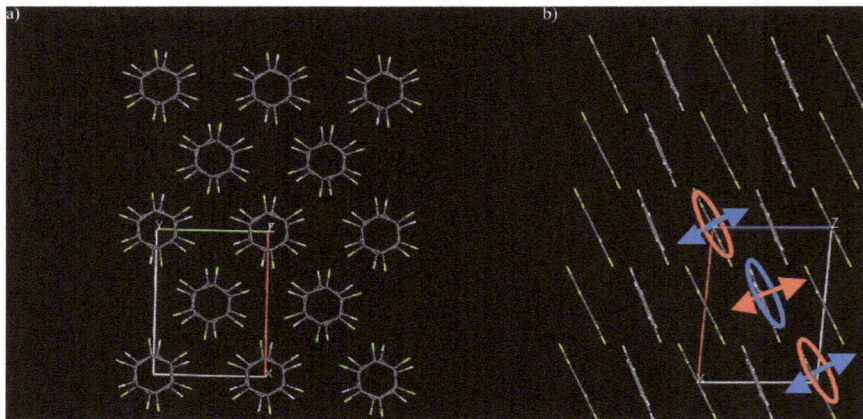

**Figure 2.4.** Two views of the structure of the lowest temperature phase (IV) of benzene:hexafluorobenzene. The structure is characterized by: (a) the stacking of the aromatic molecules into close-packed columns, and (b) the columns are made up of alternating molecules arranged parallel to each other. The negative electric quadrupole moment of benzene is represented, as in figures 2.2 and 2.3, and the positive electric quadrupole moment of hexafluorobenzene is represented by a blue arrow representing the concentration of positive charge along the normal to the aromatic plane and a red circle corresponding to the depletion of negative charge around the plane of carbon atoms. The x-ray structural data used to construct this image was taken from the Cambridge Structural Database, and are the work of Williams *et al* [5]. (Image with thanks from https://www.researchgate.net/figure/46271611_fig2_Figure-4-Two-distinct-snapshots-showing-the-crystalline-structure-of-the-equimolar).

structure, only close-packed columns of alternating benzene and hexafluorobenzene molecules.

Consequently, we may rationalize the observed arrangements of molecules in solid benzene and in solid benzene:hexafluorobenzene using simple arguments about the best way of the packing molecular quadrupole moments; quadrupole moments that are large and which change phase. The pure solids can be explained by interacting quadrupole moments of the same phase, and the binary mixture by interacting quadrupole moments of different phase. The structures revealed by x-ray diffraction investigations of these solids and, as we will see later in this volume a number of other quadrupolar molecules can be rationalized straightforwardly, and so simple models of intermolecular electrostatics may be constructed to represent the crystal engineering. And perhaps even to predict structures as yet undetermined.

To demonstrate this principle of rationalizing observed crystal structures by a consideration of the interaction of the first non-vanishing molecular moment, consider ferrocene; the structure ferrocene is displayed in figure 2.5.

Ferrocene (bis($\eta^5$-cyclopentadienyl)iron) is an organometallic compound with the formula $Fe(C_5H_5)_2$. It was first synthesized in 1951, and is the prototypical metallocene; that is, an organometallic compound consisting of two cyclopenta-dienyl rings bound on opposite sides of a central metal atom. X-ray crystallography revealed that the carbon–carbon bond distances are 1.40 Å within the five-membered rings, and the Fe–C bond distances are 2.04 Å. Although x-ray crystallography of the monoclinic crystal suggested the $C_5H_5$ rings were in a

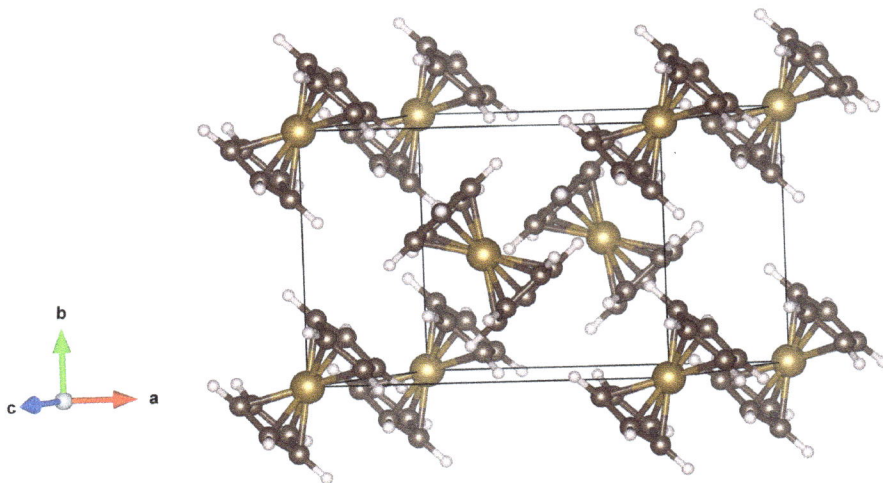

**Figure 2.5.** The structure of ferrocene. The structure clearly displays the classic 'herringbone' arrangement of molecules that is indicative of a structure arising primarily from the interaction of molecular quadrupole moments. Depending upon the relative orientation of the two organic rings in these molecules, there are a number of possible structures for this material. There are three different crystalline phases: the disordered monoclinic crystalline phase ($T > 164$ K), the metastable triclinic phase ($T < 164$ K), and the stable orthorhombic phase ($T < 250$ K). In the monoclinic phase, the unit cell has the following dimensions: $a = 10.528$ Å, $b = 7.602$ Å, $c = 5.923$ Å, with $\beta = 121.05$ °; the occupancy of the unit cell ($Z$) is 2.

staggered conformation, it has been shown that in the gas phase the organic rings are eclipsed. The staggered conformation is believed to be most stable in the condensed phase due to crystal packing.

In terms of bonding, the iron centre in the sandwich is usually assigned to the +2 oxidation state. Each cyclopentadienyl ring is then allocated a single negative charge, bringing the number of $\pi$-electrons on each ring to six, and thus making them aromatic (cf. benzene). These twelve electrons (six from each ring) are then shared with the metal via covalent bonding. When combined with the six $d$-electrons on $Fe^{2+}$, the complex attains an 18-electron configuration. Due to the delocalized bonding between the aromatic $C_5H_5$ rings and the large spherical iron atom, the aromatic $C_5H_5$ rings are free to rotate relative to each other; with a low barrier about the $C_5H_5$(centroid)–Fe–$C_5H_5$(centroid) axis.

Of interest to our discussion is the packing of the ferrocene molecules in figure 2.5. Each of the two aromatic $C_5H_5$ rings in a ferrocene molecule will have an electric quadrupole moment, of order, the same sign and magnitude as benzene (which has been borne out by experiment on solutions of ferrocene); and the presence of the electron rich iron atom between the two aromatic rings will further increase the size of the quadrupole moment of the ferrocene molecule. Figure 2.5 clearly shows that this molecule crystallizes in a manner similar to that seen in benzene, rather than that seen in benzene:hexafluorobenzene. That is, the crystal is composed of parallel and perpendicular pairs of molecules; the herringbone structure seen in figure 2.3. We note, that the strength of the intermolecular potentials in ferrocene is far stronger than that to be found in solid benzene; whereas benzene melts at 5.5 °C, ferrocene melts at 172.5 °C.


## 2.2 Self-assembly

Imagine a collection of atoms or molecules in the gaseous state. They are at a temperature that gives to each atomic, or molecular entity more energy than is required to separate the species when they are bound together by non-bonding intermolecular interactions in a stable solid. The molecules or atoms are colliding with each other at high velocities ($v$); velocities given by their thermal kinetic energy (KE), $\frac{1}{2}mv^2 = k_BT$, where $T$ is the temperature, $k_B$ is the Boltzmann constant and $m$ the mass of the particle. The particles have too much energy to form bonds between themselves and so cannot condense; that is, the kinetic energy of the gas molecules exceeds the attractive potential energy (PE) represented by the attractive interactions between these molecules. They behave as classical particles and bounce off each other like billiard balls.

In a gas, at relatively low temperatures (defined by the ratio KE:PE) one of the most important contributions to the PE between molecules or atoms is the attractive electrostatic force arising from the long-range intermolecular forces, which varies as $r^{-6}$, where $r$ is the separation of the particles. Figure 2.6 shows the form of the attractive electromagnetic force between two isolated atoms as a function of their separation, $r$.

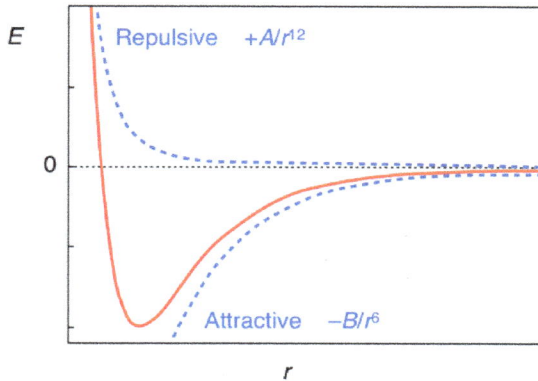

**Figure 2.6.** Modelling the interactions of molecules.

A schematic representation (in red) of the Lennard-Jones or 6–12 potential introduced in 1924 by John Lennard-Jones to model the electromagnetic-force between two atoms. The horizontal axis defines the distance ($r$) between the two atoms, and the vertical axis defines the potential energy of interaction ($E$) as a function of distance. At long-range, there is an attractive force of interaction ($E(r)$ falls as the distance between the molecules falls) scaling as $B/r^6$, as shown by the lower dashed blue curve, but at short-range the electrons of the two atoms start to repel strongly and the interaction rapidly becomes repulsive scaling as $A/r^{12}$, as shown by the dashed blue curve ($A$ and $B$ are two sample-dependent constants). The point of balance or equilibrium is given by the energy minimum seen in red.

Thus, atoms and molecules will always attract each other at large distances, hence solids and liquids exist, but these interacting species repel each other when their electrons interact at short distances, hence densities remain finite. This simple graph explains the origin and nature of the condensed phases, and of the sense of touch and of pain. This curve also explains why the densities of all solids are in relatively narrow range, of order, 0.916 g cm$^{-3}$ for ice and 22.5 g cm$^{-3}$ for osmium.

But of interest here is what happens when the PE of the interacting cloud of atoms, or molecules becomes sufficiently large to dominate the available thermal KE? That is, how do atoms and molecules condense? Alternatively, we could ask: is the manner of the condensation of a cloud of Na$^+$ ions and F$^-$ ions the same as for a cloud of neon atoms, and what governs the form of the resulting solid-state architectures?

Of course, the atom is merely a collection of electrical charges subject only to the forces generated by other, neighbouring electrical charges, or an ensemble of uncharged spherical atoms interacting via much weaker dispersion forces. In much the same way that we organize ourselves in our society, where we are subject to social forces or pressures. So, as the atoms and molecules come closer together as the available thermal energy (the force of chaos) falls, they are increasingly subject to the electrostatic forces of neighbouring molecules. These intermolecular forces can be very large, as in the case of the condensation at high temperatures (the boiling temperature of sodium fluoride is 1695 °C) of ions to form ionic solids, or the condensation of helium atoms at low temperatures (the boiling point of helium is −246 °C). Indeed, the temperature of this condensation is an expression of the strength of the intermolecular, or interatomic interaction, as the attractive intermolecular potential energy is trying to bring about condensation, but the thermal kinetic energy is generating motion and so disrupts condensation.

What happens can best be described as saying that the gaseous ions, or the neon atoms self-assemble to form an arrangement that is the most stable in the environment they find themselves to be in (the pressure, the temperature, the presence of any applied electric or magnetic fields). It is always the ratio of the molecules' KE and its PE that determines at which temperature the condensate forms. The final architecture of the crystal will be determined by the orientational dependence of the intermolecular forces once the molecules or atoms are close, but not before their relative positions are frozen. For example, as our cloud of sodium cations and fluoride anions are able to interact strongly, the powerful electrostatic forces that exist between them are not so easily disrupted by the KE of the ions and the electrostatic attraction between the ions leads to a close-packed structure, which is so ordered as to maximize attractive forces (lots of anions next to cations, and vice versa), but to minimize repulsive forces (lots of anions near other anions, or cations near other cations). The not yet bonded ions, are on a multi-dimensional potential energy surface that leads them naturally (from a consideration of the balance of attractive and repulsive forces acting upon them) to the close-packed cubic structure. The same can be said for the condensing of neon atoms, but here the dispersion forces are only important when the available KE has fallen a lot further.

The same arguments apply to the self-assembling of benzene molecules for a vapour of benzene. Here it is the first non-vanishing electrical moment that determines the packing in the condensed phases, and the two lowest energy (highest PE) orientations of quadrupolar molecules, of a single type, are the two orientations found in solid benzene; that is, a slipped parallel and a perpendicular arrangement of any pair of molecules. It is the balance of the weak, attractive quadrupole–quadrupole electrostatic interaction and repulsive orientations of the benzene molecules that leads them to the observed structure.

If the condensed state is susceptible to thermal fluctuation, which manifest themselves as low-frequency vibrations of the weak intermolecular bonds between the molecules that compose the solid, and of low frequency vibrations of the molecules themselves, then it is likely that there could well be structural changes at well-defined temperatures; that is, phase transitions. Here, the crystal structure changes to generate a new structure that is stabler in the new thermal environment. Again, the stability (and hence the structure) are determined by the balance of the attractive and repulsive forces acting on the target molecule; that is, on its potential energy surface in the thermally activated crystalline environment. Indeed, it may be possible to engineer the crystal so as to minimize the effect of such disruptive thermally-driven fluctuations, or low-energy intermolecular vibrations.

One of the biggest questions in crystal engineering, and one which is of great interest to the pharmaceutical industry is the manner in which crystals form from saturated solutions or from the molten state; that is, the kinetics of self-assembly.

The building of a macroscopic crystal needs to be considered in a stepwise manner, and these steps may be quite discrete:

one molecule → few molecules → many molecules → nucleus of a crystal →

crystal.

This sequence need not be a simple and continuous one. A mid-size cluster may be formed and then be unable to grow further so that it re-dissolves and an alternate pathway for nucleation needs to be found. A classic example is provided by acetic acid, in which nearly 90% of the liquid consists of a hydrogen-bonded dimer seen in the schematic below, but these dimers are not seen in x-ray studies; what are seen are catemers (shown below). One can say that the dimer may form easily but is unable to grow further because very weak methyl–methyl interactions would have to be the main cohesive interactions in the further assembly of dimers. The catemer is preferred because its formation provides a pathway for crystal growth quite readily in at least one dimension, and other hydrogen bonds could then form at right-angles to this 1-dimensional chain. We do not know, in a general sense, how molecular crystals are built up. Are small clusters formed which increase in an orderly way to give larger clusters, or are the events more irregular?

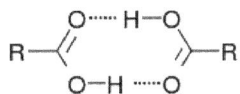

dimer

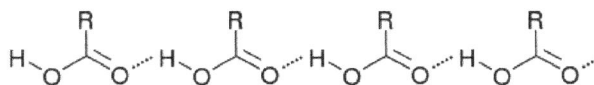

catemer

## Further reading

[1] Buckingham A D 1967 *Adv. Chem. Phys.* **12** 107–42
[2] Bhagavantam S and Suryanarayana D 1949 *Acta Crystalogr.* **2** 21;
Jahn H A 1949 *Acta Crystalogr.* **2** 30
[3] Battaglia M R, Buckingham A D and Williams J H 1981 *Chem. Phys. Lett.* **78** 421–3
[4] Cox E G, Cruickshank D W J and Smith J A S 1958 *Proc. R. Soc.* A **247** 1
[5] Williams J H, Cockcroft J K and Fitch A N 1992 *Angew. Chem. Int. Ed. Engl.* **31** 1655–7

Crystal Engineering
How molecules build solids
**Jeffrey H Williams**

# Chapter 3

# The Classification of crystals

Form is a diagram of forces

(D'Arcy Wentworth Thompson, 1860–1948)

## 3.1 The intimacy within solids

Solids are not *lifeless* permanent structures where the component molecules and atoms are held at certain distances and orientations giving a 3-dimensional structure that can exist for eternity. In ice, for example, at temperature close to its melting point there is sufficient motion that the networks of hydrogen bonds that hold the crystal together, are forming and breaking at $10^5$ per second. Yet this complex dynamic allows ice to have a structure that exists over extended distances (Antarctica), yet in water it only permits a short-range structure, extending over a few molecular diameters.

But because of the manner in which the various sub-units or individual molecules that compose the solid are bound together, they are able to undergo concerted or coordinated motion where all the component molecules move together. These concerted or coherent motions give solids their particular properties, and allows for the possibility of structural instabilities. That is, at a particular temperature, the thermal excitation of the solid (a solid absorbs thermal energy by the excitation of the weak bonds between the constituent molecules) arising from the available thermal energy causes the existing 3-dimensional solid structure to become unstable and the solid may undergo a structural phase transition where the 3-dimensional solid structure changes; it rearranges itself in an instant. In a solid-state phase transition, the material remains a solid and the chemical formula does not change, but all the sub-units in the solid are organized differently so as to maximize the intermolecular forces in the new, more thermally agitated (or less thermally agitated if we are lowering the temperature of the solid) environment.

doi:10.1088/978-1-6817-4625-8ch3

It is the thermally-driven motion of atoms and molecules that give all the phases of matter their strange and unique properties. There is motion everywhere we look in Nature.

## 3.2 Crystallography

We now need to consider how to describe and to classify the solids formed from molecules; from the tens of millions of molecules of which we presently aware.

*When we examine a crystal, particularly a large well-formed crystal, the quality and geometric perfection of the sides pleases us. The perfection and sharpness of the angles doubles our pleasure. Then we turn the crystal and upon viewing a second side, which is in all respects identical to the first side, the pleasure of observation seems to be squared. Then looking at a third side, a fourth side, our pleasure appears to increase in orders of magnitude.*[1]

Crystals may only be solid manifestations of ionized atoms or neutral molecules being held together at particular distances and at particular orientations by intermolecular forces of an electromagnetic origin, yet they are as near to an aesthetic perfection as we may come across in Nature.

All crystals contain a great deal of symmetry and order, but this is particularly the case with ionic crystals. And it is via this inherent internal long-range order, which is often reflected in the shape and form of the crystal one can hold in one's hand that a classification of crystals has arisen. Crystals are defined by *symmetry operations*; for example, does rotation of the crystal about some axis, or reflection through some plane in the crystal result in the same structure being viewed (that is, bringing itself into co-incidence), or does the symmetry operation generate a new structure? Let us take the example of sea-salt or sodium chloride, which crystallizes into cubic forms.

How can we define a cubic crystal? Well, if you rotate a cube (which is a very symmetric object) by 90° about an axis passing through the centres of a pair of opposite faces, you return the cube to its original position; the two positions may be superimposed, because nothing has changed by the operation of rotation. For a cube, such a rotation is said to be a *symmetric operation*; a $C_4$ operation as there are four such axes of rotation.

Similarly, if we slice a cube in half perpendicular to a $C_4$ axis of rotation, the two half-cubes are mirror images. The plane of the slice is then said to be a *mirror plane* or a *plane of symmetry* and is denoted by $m$. In a cube there are three such planes (one of which will slice through opposing corners of the cube). A cube also has other symmetry axes: six 2-fold axes ($C_2$ or 180° rotators) passing though the midpoints of opposite edges, and four 3-fold axes ($C_3$ or 120° rotators) passing though opposite corners. Besides its symmetry axes, a cube has other symmetry elements. The cube also has the important symmetry property, an *inversion centre and centre of*

---

[1] This is almost an intuitive derivation of Ludwig Boltzmann's expression for the entropy of a system, $S$, i.e. $S = k_B \log W$, where $W$ is a measure of the number of states there are in that system. Quotation taken from *The Rationale of Verse* 1843.

*symmetry*, *i*; such that any point that outlines the cube when projected through *i* comes into coincidence with itself.

In this way, we may describe the symmetry of the cube in terms of its symmetry elements or symmetry operations (they are mathematical operations): three $C_4$, four $C_3$, six $C_2$, three *m* perpendicular to $C_4$, six *m* perpendicular to $C_2$, and *i*. More than one symmetry operation can belong to a given symmetry element. So $C_4$ can transform the symmetry operations of rotation by 90°, 180°, 270° and 360°; that is, $C_4(\pi/2)$, $C_4(\pi)$, $C_4(3\pi/2)$ and $C_4(2\pi)^2$.

For all crystals, the set of all symmetry operators forms a *group*; and this group is a mathematical entity which obeys certain rules. These symmetry operations in crystals are entirely analogous with the application of symmetry in the quantum mechanics of sub-atomic particles, with a view to determining which particular forces of Nature act upon which sub-atomic particles. Indeed, symmetry is as powerful a tool and a means of studying problems in physics as it is in art.

The cubic system contains a great deal of symmetry. If there is less symmetry in the system being examined, for example, if the $C_4$ axes are actually found to be only $C_2$ axes, and there are no $C_2$ axes with mirror planes normal to them, then we are looking at a system of lower symmetry (in fact, the point group of iron pyrites defined as $T_h$) than the full cubic system seen in rock-salt. As any artist will tell you, symmetry is at the heart of looking at and understanding Nature. And even at the molecular scale, symmetry is the key to describing structure and function. Molecules often interact symmetrically, producing crystals which therefore possess something of the symmetry of the molecule. Textbooks often deal with this subject in a manner remote from practical use. This subject, which is essentially about the manipulation of objects in three dimensions, is highly mathematical. Indeed, the physics and mathematics of crystallography was established in the 18th and 19th centuries, well before we had the slightest idea of how atoms and molecules interact and how they are arranged in crystals.

The faces of a crystal (usually, the first means of trying to identify to which symmetry class a crystal belongs) can be described by a set of three non-coplanar axes. If we construct such axes so that their lengths are in definite ratios, a finite set of crystal systems can be identified, even though the total number of crystals is enormous. Consider a non-coplanar axis system (*a*, *b*, *c*) starting at an origin *O*. These three axes may be cut by a plane ABC, making intercepts at *OA*, *OB*, *OC*. Expressed as fractions of the crystal axes, their lengths are *OA/a*, *OB/b* and *OC/c*; and they have reciprocals *a/OA*, *b/OB* and *c/OC*. Given that ratios of rational numbers can always be reduced to whole numbers by use of appropriate multiplicative factors (the law of rational intercepts), these reciprocal intercepts will always be in the ratio of the whole numbers (*hkl*) for planes that are crystal faces.

In 1839, the Welsh mineralogist and physicist William Hallowes Miller (1801–1880) introduced such reciprocal intercepts to designate crystal faces; they can also be used to define planes within crystals. If a face is parallel to a given axis, the

---

[2] These symmetry elements, or operations of a cube may be seen at https://sites.google.com/site/greatmathmoments/group.

intercept is at infinity, and the corresponding Miller index will be $1/\infty = 0$. Depending upon the axes required for the law of rational intercepts to hold, all the millions of crystals can be divided into only seven crystal systems.

In crystallography, the terms crystal system, crystal family and lattice system each refer to one of several classes of space groups, lattices, point groups or crystals. They are similar but slightly different and are all defined by symmetry operations; consequently, there is often confusion between them: in particular the trigonal crystal system is often confused with the rhombohedral lattice system, and the term *crystal system* is sometimes used to mean *lattice system* or *crystal family*. Informally, two crystals are in the same crystal system if they have similar symmetries; though there are many exceptions to this *rule*.

Space groups and crystals are divided into seven crystal systems according to their symmetry properties (point groups), and into seven lattice systems according to their Bravais lattices[3]. Five of the crystal systems are essentially the same as five of the lattice systems, but the hexagonal and trigonal crystal systems differ from the hexagonal and rhombohedral lattice systems. The six crystal families are formed by combining the hexagonal and trigonal crystal systems into one hexagonal family, in order to eliminate confusion. The 14 Bravais lattices are grouped into seven lattice systems: triclinic, monoclinic, orthorhombic, tetragonal, rhombohedral, hexagonal and cubic.

In a crystal system, a set of point groups and their corresponding space groups are assigned to a lattice system. Of the 32 point groups that exist in three dimensions, most are assigned to only one lattice system, in which case both the crystal and lattice systems have the same name. However, five point groups are assigned to two lattice systems, rhombohedral and hexagonal, because both exhibit 3-fold rotational symmetry. These point groups are assigned to the trigonal crystal system. In total there are seven crystal systems: triclinic, monoclinic, orthorhombic, tetragonal, trigonal, hexagonal and cubic.

The relation between 3-dimensional crystal families, crystal systems and lattice systems is given in table 3.1.

The seven crystal systems consist of 32 crystal classes (corresponding to the 32 crystallographic point groups). The names given to the 32 crystal classes were formulated in the 19th century, and are based on the morphology of the crystal of the material in question. Thus, in the monoclinic system there are three crystal classes: monoclinic-sphenoidal (Schönflies symmetry $C_2$), monoclinic-domatic (Schönflies symmetry $C_s$) and monoclinic-prismatic (Schönflies symmetry $C_{2h}$); and these names define the shape of a specimen of the material. These names provide no real information about the arrangement of atoms in the crystals as they define the outward appearance of a macroscopic crystal, as the physicists who derived these names knew nothing of molecules and polyatomic assemblages. However, there will be an inference of outward morphology on internal atomic structure.

---

[3] Named after the French crystallographer, geographer and physicist Auguste Bravais (1811–1863), who was one of the first to apply physics to understanding the form of the landscape.

**Table 3.1.** Relationship of crystal classifications.

| Crystal family | Crystal system | Required symmetries of point group | Point groups | Space groups | Bravais lattices | Lattice system |
|---|---|---|---|---|---|---|
| Triclinic | Triclinic | None | 2 | 2 | 1 | Triclinic |
| Monoclinic | Monoclinic | 1 2-fold axis of rotation or 1 mirror plane | 3 | 13 | 2 | Monoclinic |
| Orthorhombic | Orthorhombic | 3 2-fold axes of rotation or 1 2-fold axis of rotation and 2 mirror planes. | 3 | 59 | 4 | Orthorhombic |
| Tetragonal | Tetragonal | 1 4-fold axis of rotation | 7 | 68 | 2 | Tetragonal |
| Hexagonal | Trigonal | 1 3-fold axis of rotation | 5 | 7 | 1 | Rhombohedral |
| | | | | 18 | | |
| | Hexagonal | 1 6-fold axis of rotation | 7 | 27 | 1 | Hexagonal |
| Cubic | Cubic | 4 3-fold axes of rotation | 5 | 36 | 3 | Cubic |
| **6** | **7** | **Total** | **32** | **230** | **14** | **7** |

**Table 3.2.** The geometry of crystal systems and Bravais lattices.

| Crystal system | Axes | Angles | Bravais Lattices |
|---|---|---|---|
| Cubic | $a = b = c$ | $\alpha = \beta = \gamma = 90°$ | (1) Simple<br>(2) Face-centred<br>(3) Body-centred |
| Tetragonal | $a = b; c$ | $\alpha = \beta = \gamma = 90°$ | (4) Simple<br>(5) Body-centred |
| Orthorhombic | $a; b; c$ | $\alpha = \beta = \gamma = 90°$ | (6) Simple<br>(7) End-centred<br>(8) Face-centred<br>(9) Body-centred |
| Rhombohedral | $a = b = c$ | $\alpha = \beta = \gamma$ | (10) Simple |
| Hexagonal | $a = b; c$ | $\alpha = \beta = 90°; \gamma = 120°$ | (11) Simple |
| Monoclinic | $a; b; c$ | $\alpha = \gamma = 90°; \beta$ | (12) Simple<br>(13) Face-centred |
| Triclinic | $a; b; c$ | $\alpha; \beta; \gamma$ | (14) Simple |

All crystalline materials must fit into one of the symmetries described in table 3.1, which does not apply to quasicrystals.

In table 3.2 we see details of the geometry of the Bravais lattices. Note that in the Miller index for a crystal face, it is only the ratio $h{:}k{:}l$ that is of importance; thus, (220) would be the same face as (110); the parentheses denote a crystal face[4]. However, for planes within a crystal, multiplication by an integer would change the inter-planar spacing; thus, the planes 200 would include the 100 and the set of planes midway between them. (Miller indices without parentheses denote a set of planes.) If we wish to refer to all equivalent planes of a crystal; that is, a *form* of a crystal, we use curly brackets; for example, we would say the cubic sodium chloride crystal has the {100} form.

Tables 3.1 and 3.2, display essential characteristics of crystal systems; that there are only $C_2$, $C_3$, $C_4$ and $C_6$ rotational axes to be found in Nature. There are, for example, no 5-fold axes, or 8-fold axes. There are, however, molecules that have a 5-fold symmetry; for example, ferrocene $Fe(C_5H_5)_2$ (see figure 2.5). But this molecular symmetry is not necessarily transformed into the symmetry of the crystals formed from those molecules. The reason for this incommensurability is that it is not possible to fill Euclidian space with objects of 5-fold symmetry. The absence in crystals of axial symmetry elements other than 2-, 3-, 4- and 6-fold axes tells us that crystals can only be constructed of repeat subunits which fill space in distinct

---

[4] With *hkl* values, there are five different notations. The diffraction plane indices *hkl* have no brackets. Indices of a crystal face are given the symbol (*hkl*) and for all symmetry equivalent faces (e.g. as for cubic symmetry) are given the symbol {*hkl*} (both are real space usage). So {100} for cubic symmetry comprises (100), (010), (001), (−100), (0−10), and (00−1). To indicate a lattice direction, the notation changes to [*uvw*] (where $d^* = ua^* + vb^* + wc^*$) where *uvw* are integer values *hkl* for Bragg diffraction; for all symmetry equivalent directions this becomes <*uvw*> (both are reciprocal space usage).

geometric patterns. That is, the crystal forms seen in Nature are outward manifestations of internal regularities and structure.

It is upon a lattice that one may describe in space a regular array of subunits. Such a lattice is a regular, infinite array of points in space, arranged at the intersection points of three ranges of equidistant planes. These points may be joined together so as to construct unit cells, which fill all space. As we have seen, there are restrictions on the forms of unit cells that fill all space. Sometimes we do not use a unit cell in which all the points are at the corners, but instead use a face-centred, end-centred, or body-centred unit cell. Unit cells in which all the points are at the corners form primitive lattices, the others form centred lattices. But there are only 14 possible lattices; they were identified by Auguste Bravais and are termed the Bravais lattices (see tables 3.1 and 3.2).

A lattice is merely a mathematically useful array of points in space. If at each of the points in a lattice we were to place an atom or a molecule, held in place by the balance of the attractive and repulsive intermolecular forces, we would obtain a crystal structure. All crystal structures are based on one or other of the 14 Bravais lattices. Just as a lattice is composed of unit cells that pack together to fill space, a crystal structure is also made of unit cells. So, if we know the location of the atoms, or ions, or molecules within a unit cell we have enough data to determine the crystal structure.

The unit cell provides a convenient way to locate the centres of mass of each atom, ion or molecule within a crystal structure. The lengths of the sides of the unit cell, $a$, $b$, $c$, are taken as unit lengths, and the position of any point in the cell is designated as ($u = x/a$, $v = y/b$, $w = z/c$); for example, in the cubic unit cell, the body-centred position would be ($\frac{1}{2}$ $\frac{1}{2}$ $\frac{1}{2}$), a face-centred position ($\frac{1}{2}$ 0 $\frac{1}{2}$), and so on. A general position would be, $uvw$. In terms of the dimensions of the unit cell, we may write for the distance between sets of planes $hkl$ (in the cubic system):

$$d_{hkl} = a_0/\sqrt{(h^2 + k^2 + l^2)}, \tag{3.1}$$

where $a_0$ is the edge of the unit cell. For example, the distance between the 111 planes in rock-salt is $d_{111} = a_0/\sqrt{3} = 3.25$ Å.

Solids and crystals are composed of near-infinite arrays of atoms, or molecules, or ions, yet how and why do they fill space in the manner they do? This is the great question of solid-state science; for example, why is the crystal structure of calcite the way it is, yet the crystal structure of rock-salt is totally different? It is similar, yet very different, with different physical properties.

Space groups are essential to crystallographers and those seeking to investigate how crystals are engineered. Firstly, they allow crystal structures to be described without the need to list every atom and its 3-dimensional coordinates within the unit cell. Secondly, determination of space-group symmetry simplifies the determination of a crystal structure, whether from single-crystal or powder diffraction data. Finally, the coordinates of a crystal structure cannot be refined reliably without a knowledge of the space-group symmetry. However, it is not practical, nor is it necessary to know all of the space groups as the frequency of space groups, as observed in the Cambridge Structural Database, provide the following statistics:

| Crystal System | Frequency in the Database |
| --- | --- |
| Triclinic | 21% |
| Monoclinic | 53%[a] |
| Orthorhombic | 21% |
| Tetragonal | 2% |
| Trigonal | 2% |
| Hexagonal | 0.5% |
| Cubic | 0.5% |

[a] And P2(1)/c is by far and away the most common space group in the monoclinic system for packing non-chiral molecules with no symmetry; they will pack best with a screw axis perpendicular to a glide plane (based on the ideas of Alexander Kitaigorodsky) as these symmetry elements avoid the formation of void holes in the crystal structure, so enabling close packing (see http://pd.chem.ucl.ac.uk/pdnn/symm3/sg14a1.htm for technical details). According to the Cambridge Structural Database, this space group accounted for around 1 in 3 of all crystal structures in this system. For chiral molecules one loses the inversion centre and glide plane so space group P2(1) becomes the most common space group.

These statistics demonstrate that molecular crystals usually have low-symmetry crystal structures; over 95% of the structures belong to either the triclinic, mono-clinic, or orthorhombic crystal systems. The great variety of molecular shapes are accommodated in only three crystal systems.

The crystal system to which a solid belongs will immediately tell us something about the manner in which the atoms and molecules that compose the crystal are arranged within the solid. However, to see precisely how the individual atoms and molecules are ordered and arranged in the crystal, we need to look into that solid. But how does one explore this internal perfection of the crystal, and determine the orientation and spacing of the atoms and molecules within?

## 3.3 X-ray diffraction

In 1913, the German physicist Max von Laue (1879–1960) suggested the possibility of using the recently discovered x-rays as a means of probing the structure of crystals; a suggestion for which he won the Nobel Prize for physics of 1914. He reasoned that as the wavelength of x-rays (typically 1–2 Å) is of the same order as the inter-atomic distances inside the crystal, this radiation was ideal for the study of crystal structure. If x-rays were allowed to strike the crystal, the rays would penetrate into the crystal and would be scattered or reflected by the electrons of the atoms or ions that form the crystal. The x-rays reflected from different layers of the atoms would then undergo interference to produce a diffraction pattern. In other words, crystals would act as a 3-dimensional grating for x-rays[5].

---

[5] This scattering process is entirely different from the reflection of visible light from a polished surface, a property which is due to the change in velocity of light at the interface.

Consider a beam of x-rays, of wavelength $\lambda$ incident upon the face of a crystal. Let us further suppose that the crystal face is made up of a regular array of atoms located at lattice points. The electric field of the x-ray photons will interact with the electrons in the atoms of the array, and as a result, the x-rays will be scattered from each atom. These scattered rays or photons do not form a coherent beam as they will interfere destructively with each other. First consider the scattering by a single line of equispaced scattering centres of separation $a$. The condition that waves scattered from two adjacent points be in phase is that their path difference equals an integer number (n) of wavelengths. For a line of equispaced scattering centres, the path difference is:

$$t - s = a \cos \varphi - a \cos \theta = n\lambda,$$

where $t - s$ is the path difference of two wavelets whose angle of incidence upon the array and scattering from the array are given by $\theta$ and $\varphi$. When $n = 0$, we have the zero-order diffracted rays forming a cone generated around the direction of incidence. The direction of this scattering cone depends only upon the direction of incidence of the incoming x-rays and not upon the spacing of the scattering centres. When $n$ becomes finite, the higher order diffracted beams are also cones, but their position now depends on the spacing, $a$. It was the observation that a randomly ordered set of scattering centres (an amorphous solid) generated diffuse scattering that led to the explanation for the scattering from a 3-dimensional crystal.

Two British physicists, William Lawrence Bragg (1890–1971) and his father William Henry Bragg (1862–1940) determined the first crystal-structure by examining sea-salt; that is, they demonstrated the cubic structure of the NaCl crystal using x-rays. In the Braggs' treatment, the x-rays strike the crystal at a defined angle $\theta$; these rays penetrate the crystal, and are reflected by different parallel layers of ions in the crystal.

A strongly diffracted beam will only be seen if all the rays reflected from the various layers of the crystal are in phase, which is why the wavelengths have to be of the same size as the spacings[6]. The waves or rays reflected by different layers will be in phase if the difference in the path length of the waves reflected from the successive planes is equal to an integral number of wavelengths (see figure 3.1)

Figure 3.1 tells us that the beams of x-rays reflected from deeper layers travel further to reach the detector. Three, in-phase x-rays are shown approaching the crystal (rays 1, 2 and 3); one wave is reflected from the first layer of atoms while the second wave is reflected from the second layer of atoms. The wave reflected from the second layer travels a greater distance before emerging from the crystal than the first wave. The extra distance travelled is given by the solid lines between the dashed lines within the crystal. For constructive interference to take place the extra distance travelled by the more penetrating beam must be an integral multiple of the wavelength ($\lambda$) of the x-ray radiation. Analysis of this set-up gives us Bragg's Law, or the Bragg condition (W L Bragg) of 1912 for the angular dependence of the

---

[6] Because of the very different length scales of visible light and inter-atomic distances within a crystal, trying to use visible light to study the internal structure of crystals would be like trying to use a bulldozer to sort golf balls; the diffraction and interference of the visible light would be difficult to observe due to the incommensurate length scales between the inside of the crystal and the incoming probing radiation.

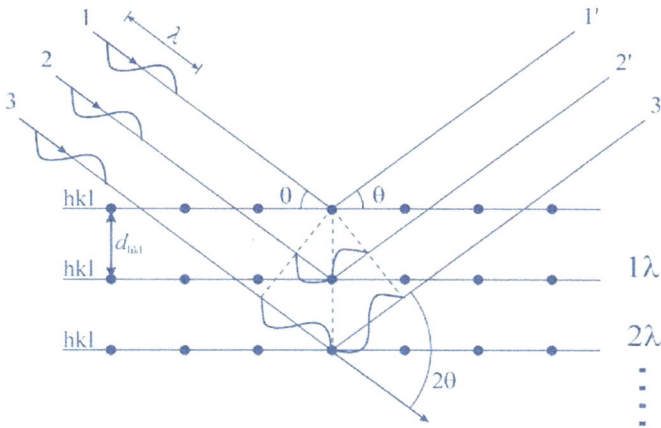

**Figure 3.1.** X-ray reflection and interference from crystals.

scattering; $n\lambda = 2d \sin \theta$, where $n$ is a number and $d$ is the inter-atomic spacing (the distance between two atomic layers of the crystal).

Often it is not possible to obtain a large single, perfect crystal upon which to undertake x-ray scattering experiments, and one must make do with powders of the material. In such powder diffraction studies, instead of a single crystal with a face (*hkl*) at a well-defined angle $\theta$ to the incoming x-ray beam, a mass of finely divided crystals is placed in the beam. In such an experiment, there are so many tiny crystals in the sample that for any particular set of planes (*hkl*), which meet the Bragg condition, $\lambda = 2d_{hkl} \sin \theta_{hkl}$, there is always an appropriately ordered set of crystals. And the direction of the scattered beams is determined only by the angle $\theta_{hkl}$. So, for each set of planes (*hkl*), the scattered radiation forms a cone; and if a flat plate is used as a detector the result will be a series of rings around the incident beam position with the scattering cones (or lines).

To *index* a powder diffraction pattern, one measures the distance of each line from the central beam spot to obtain the Bragg angle corresponding to each line of the powder pattern. Then from $d_{hkl} = (\lambda/2)/\sin \theta$ we may calculate the spacing for each set of reflecting planes (*hkl*). To index the lines; that is, to assign each one to the proper set of indices *hkl*, we compare these experimental spacings with theoretical spacings. To do this, we must know the value of $a_0$, but this can be found by fitting two or three experimental $d_{hkl}$ to the appropriate theoretical formula (depending on the crystal class).

Figure 3.2 displays some typical powder diffraction patterns. Here, the sample is the binary adduct formed by mixing equimolar quantities of benzene and hexafluorobenzene (we will discuss this material later). The sample is not a single crystal of this material, but a multitude of randomly oriented crystallites; that is, a powder. From the data in figure 3.2, this material can be seen to undergo three solid-state phase transitions over the temperature range examined. The problem that must be solved is how to go from a measured set of data, as in figure 3.2, to the crystal structure of this material; that is, how does one determine the architecture of the

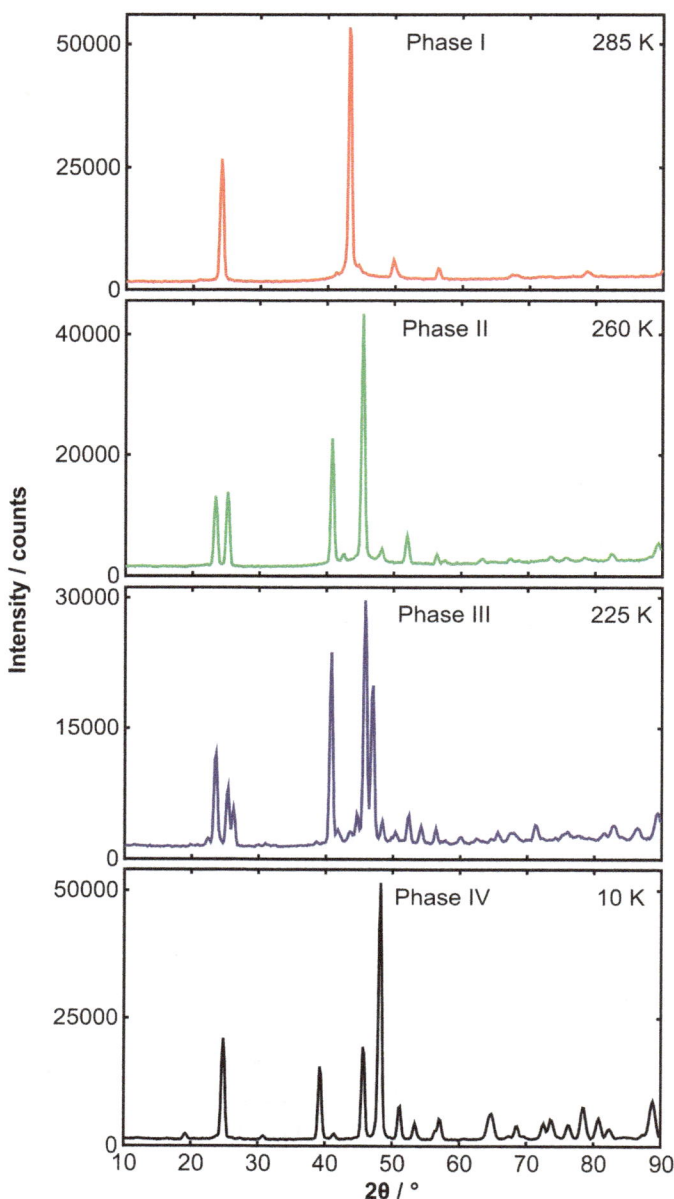

**Figure 3.2.** Some high-quality neutron diffraction data. These four diffraction patterns are of the same material, the binary adduct benzene:hexafluorobenzene, but measured at four different temperatures; see image for details. The meaurements were made on the neutron diffractometer, D1B, at the Institut Laue-Langevin, Grenoble, using neutrons of wavelength 2.52 Å. The enormous differences between these diffraction patterns arises because of the structural phase transitions seen in this material at 205 K, 247.5 K and 275 K. What is also clearly seen in the highest temperature measurement is the broad thermal diffuse scattering (from large amplitude motion of the scatterers) underlying the Bragg lines at the highest temperature. In phase IV, the small peak at 19° is the (001) reflection, the large peak at 24.5° is the (110) reflection, the peak at 39.5° is the (002) reflection and the biggest line at 49° is the (−202) reflection.

solid, or how each of the molecules of benzene and hexafluorobenzene are arranged and oriented in the solid?

This is a classic problem of indirect methods in experimental science. The diffraction experiment does not measure a single quantity, which may be equated with a single unknown in a theory. In diffraction measurements, one has measured a vast amount of data, which is in some way related to the 3-dimensional coordinates of two polyatomic molecules; to the geometric organization of 24 atoms per molecular pair.

Of course, the experimenter knows that the solid is composed of two molecules, and that these two molecules retain their identity in the organic solid, and the structure of the benzene and hexafluorobenzene are well-known; consequently, one begins to search for the presence of these known molecules—the two molecules both contain a hexagon of carbon atoms. This is done rather laboriously by calculating what the diffraction pattern would be for a series of known orientations of the two molecules under consideration, and then making a comparison with the observed data. This calculation using Bragg's law is rendered trivial by the use of computers, but there is no subtle inference at work here, merely a process of calculation and comparison.

Eventually, one will arrive at a situation where the theoretical model of what you think the structure of the solid is generates a diffraction pattern that resembles in some way the measured data. Then you ask the software to fit (usually with a least-squares algorithm) the model to the data, having fixed several of the variables. In this way, one attempts to refine the model structure; for example, by varying the separation of the two hexagonal rings, or the various angles that define the orientation of the two molecules. This process of structure determination will eventually generate a model that *fits* the data. Then the structure may be said to have been solved. After having fitted the data, you will have a set of coordinates for each of the atoms in the unit cell of the solid, which via the rules of crystallography gives the structure of the entire solid.

Throughout the remainder of this volume, you will see many dozens of structures of solids, each of which has been generated by sophisticated plotting software (for example, Crystalmaker, http://www.crystalmaker.com/, or the free software Mercury, https://www.ccdc.cam.ac.uk/Community/csd-community/freemercury/) from a set of coordinates deposited in the Cambridge Structural Database (https://www.ccdc.cam.ac.uk/) in the form of a crystallographic CIF file.

It is by examining the scattering of x-rays or neutrons by crystals, and the application of the simple law due to W L Bragg that all the internal dimensions of crystals may be determined. The first crystal structure to be determined was that of sea-salt. But it was less than 40 years later, that Francis Crick and James Watson unravelled the structure of the DNA molecule by using the same technology and experimental procedure; a discovery published in April 1953 (the Nobel Prize was awarded in 1962). When one looks at the serried ranks of atoms and molecules inside the ordered world of a crystal, all marching to infinity or rather to the sharp flat face of the crystal, and when one further imagines these columns and rows of molecules all moving...breathing coherently in response to the ever-present background thermal radiation, one is reminded of Goethe's definition of architecture, *crystallized music*.

## Further reading

There are many useful articles on crystallography to be found in web-based resources such as Wikipedia, but a good source of technical detail, particularly related to x-ray diffraction studies of powders, is Advanced Certificate in Powder Diffraction by Dr Jeremy K Cockcroft, School of Crystallography, Birkbeck College, University of London at http://pd.chem.ucl.ac.uk/pdnn/chapter.htm

Mention must also be made of a publication brought out to celebrate the 50th anniversary of the Braggs' Nobel Prize; 50 Years of x-ray Diffraction by P P Ewald (https://www.iucr.org/publ/50yearsofxraydiffraction). Ewald was one of the pioneers of x-ray diffraction and this work is a hugely informative history of the subject—and is freely available.

# Chapter 4

## Non-bonded solids

Let us begin our exploration of what holds atoms and molecules together in solids; that is, the mechanism of the crystal architecture, by looking at non-bonded interactions, as demonstrated by neon.

Neon is a member of a group of chemical elements termed the noble, or the inert gases; and these names explain everything. The elements in this group of the periodic table are all gases, even when the elements around them are metals. This tells us that the atoms of these gases do not interact strongly with other atoms. Indeed, these elements have filled shells of electrons and so are not susceptible to chemical reaction under standard laboratory conditions[1]. Neon has an electronic configuration of $1s^2$ $2s^2$ $2p^6$, on either side of neon in the periodic table are the gas fluorine (with electronic configuration $1s^2 2s^2 2p^5$) and the metal sodium (with configuration $1s^2 2s^2$ $2p^6$ $3s^1$). The electronic structure of sodium and fluorine make them extremely reactive chemically.

The atoms of neon (as with the other members of its group) are spherically symmetric, and are polarizable. Neon does condense to form a solid, but at low temperatures; the melting point of neon is 24.56 K, and x-ray crystallography has demonstrated that the atoms are packed into the solid in a cubic close-packed arrangement. Given that the crystal of neon belongs to the cubic system, the unit cell parameters are $a = b = c = 4.429$ Å, and $\alpha = \beta = \gamma = 90.0°$. The structure of solid neon is as seen in figure 4.1.

Given their inertness, the packing of the atoms of the inert gases into solids would be expected to be straightforward. In fact, the lattice arrangements existing for neon,

---

[1] These gases show extremely low chemical reactivity; consequently, only a few hundred noble gas compounds have been formed. In 1933, Linus Pauling predicted that the heavier noble gases could form compounds with fluorine and oxygen. He predicted the existence of krypton hexafluoride ($KrF_6$) and xenon hexafluoride ($XeF_6$), speculated that $XeF_8$ might exist as an unstable compound and suggested $H_2XeO_4$ could form perxenate salts. These predictions were shown to be generally accurate. Xenon compounds are the most numerous of the noble gas compounds that have been synthesised.

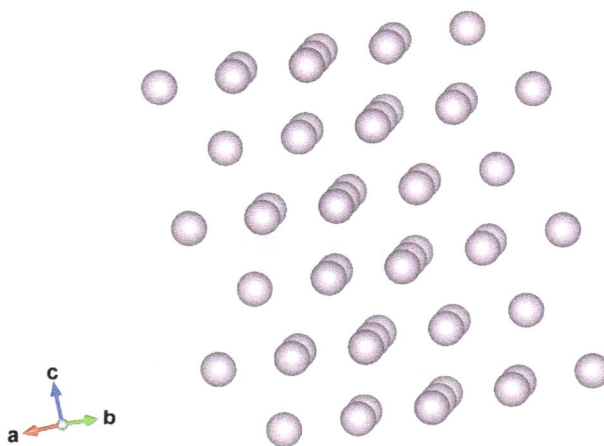

**Figure 4.1.** The structure of solid neon. It could, however, be the structure of neon, argon, krypton, or xenon, as they all form the same type of face-centred cubic lattice; only the unit cell dimensions being different; 4.429 Å, 5.25 Å, 5.721 Å and 6.197 Å, respectively. See http://www.chemtube3d.com/solidstate/_simplecubic(final).htm for an interactive animated structure for a close-packed cubic structure.

argon, krypton and xenon are not what a simple model of the packing of hard spheres would predict; but such simple modelling does predict the hexagonal close packing of helium. Solid neon, argon, krypton and xenon are all cubic, but they generate face-centred cubic structures. Evidently, there is some additional component in the intermolecular potential than a simple two-body, isotropic interaction of the van der Waals type (a simple deviation from ideal gas behaviour) would predict; and this additional component to the interatomic potential comes from three-body, or N-body dispersion interactions.

## 4.1 Dispersion interactions

In 1923, Fritz London was seeking a means of explaining the origin of the attractive electromagnetic force between two spherically symmetric helium atoms. Helium does condense, but at 4.22 K, so the minimum of the intermolecular potential for two helium atoms (figure 2.6) is very shallow. London reasoned that perhaps the electrons moving around the nucleus of each atom could couple in some way.

Fritz London went on to show that there exists, what he termed a dispersion interaction between even spherically symmetric atoms, which arises from the coupling of instantaneous fluctuations in the charge distributions of two neighbouring atoms. We do not know how electrons move in atoms, but from Earnshaw's theorem of 1839 we know that those electrons must be moving. If a fluctuation in the charge distribution on one atom gives it a fleeting electric dipole moment (the fluctuation breaks the spherical symmetry), that induced dipole will exert an electromagnetic force on a neighbouring atom and induce another electric dipole moment. Then the two dipole moments will interact to generate an attractive intermolecular potential. The presence of one atom perturbs the other atom and vice versa, and the induced attractive potential will be proportional to $1/r^6$, where $r$ is the separation of the two atoms.

This explanation is classical, but because the interaction of the two atoms in quantum mechanics is by way of the electromagnetic field of the vacuum (the quantum vacuum fluctuations), the interaction between the helium or neon atoms is actually more complex. Consider two atoms, A and B, being relatively close to each other. When the fluctuation on A arises, it generates an electromagnetic wave which is transmitted to B at the speed of light, $c$. The charge distribution responds to the wave from A, and generates a wave that returns to A. So, the interaction in quantum mechanics, is an exchange of photons between the two atoms, which takes a time $r/c$ to travel from A to B and then $r/c$ to return to A, i.e. $2r/c$. If the fluctuation on A occurs at a frequency, of order $\Delta/h$, where $h$ is Planck's constant, and if the round-trip time of the fluctuation, $2r/c$, is larger than this, the dipole on A will have fluctuated to a new position, so weakening the overall interaction. Only when the two atoms are close ($2r/c$ shorter than $\Delta/h$) is there an effectively instantaneous interaction.

Because the dispersion interaction is so weak (all interacting atoms and molecules display this intermolecular force or interaction, it is just that it is masked by other stronger non-bonding interactions in systems other than spherically symmetric atoms), the retardation of the interaction becomes so significant as to change the distance-dependence of the interaction from $1/r^6$ to $1/r^7$. Such retardation effects become significant in systems formed of colloids and macromolecules, and for measurement of the Casimir–Polder force and the Casimir effect.

What London's theory said was that there would be short-lived electric fields on each atom, which by Maxwell's equations would, however, require an expenditure of energy. To circumvent this problem of instability, London simply assumed that there existed an instantaneous electric field without energy loss, thereby ignoring Maxwell and most of the classical physics of the 19th century. Yet this assumption allowed him to go on to explain the observed condensation of gases such as helium.

The fluctuations on the neon atoms can couple to the quantum fluctuations of the vacuum and then couple back to the other neon atom (and other neighbouring neon atoms) and vice versa. In other words, matter can perturb the vacuum and be perturbed by the vacuum, so all atoms and hence all molecules would display this effect, but in helium and neon it is the easiest to describe as it is the only contribution to the intermolecular forces. The London interaction or, as it is often called today, the dispersion force is a weak intermolecular force arising purely from a quantum mechanical phenomenon.

The other two types of attractive van der Waals forces that lead to non-ideal behaviour in gases arise from classical electrostatics and the theories to explain the Keesom and Debye forces could easily have been developed in the century before the discovery of quantum mechanics; they are classical phenomena. However, the London force arises from the coupling of induced instantaneous polarization in adjacent molecules and atoms. They can therefore act between molecules and atoms without permanent polarization. London forces are the only attractive intermolecular force present between atoms of the inert gases and without these forces there would be no attractive force and they would be perpetual gases.

The strength of the dispersion interaction depends on the polarizability ($\alpha$) of the first molecule, because the instantaneous dipole moment $\mu_1$ depends on the looseness of the control that the nuclear charge of that atom exercises over the outer electrons; the cgs units of the molecular polarizability is a volume; the bigger the molecule, the more polarizable it will be. The strength of the interaction also depends on the polarizability of the second molecule as that polarizability determines how readily a dipole can be induced by another molecule. The actual quantum mechanical expression is complex involving second-order perturbation theory, but a reasonable approximation to the interaction energy, $V$, is given by the London formula. London made a Taylor series expansion of the perturbation in $1/r$, where $r$ is the distance between the nuclear centres of the interacting atoms. This expansion is known as the multipole expansion because the terms in this series can be regarded as energies of two interacting multipoles, one on each monomer.

The dispersion interaction, $V_{AB}^{(\mathrm{disp})}$, between two atoms, $A$ and $B$, may be written as

$$V_{AB}^{(\mathrm{disp})} = -C/r^6, \text{ where } C = \frac{3}{2}\frac{I_A I_B}{I_A + I_B}\alpha_A\alpha_B,$$

where $I_A$ and $I_B$ are the ionization energies of the two molecules. And as the size of the inert gas atoms increase from helium, the atomic polarizability increases and so interaction-induced polarizabilities are also generated, and these are the reason that the dispersion interactions yield crystal architectures different from that seen in helium.

## Further reading

For those readers seeking a more detailed description of non-bonded interactions, I recommend Atkins P W and Friedman R 2005 *Molecular Quantum Mechanics* 4th edn (Oxford: Oxford University Press)

# Chapter 5

## Ionic materials

In crystalline neon we observed a body-centred cubic close packing of the inert atoms resulting from weak dispersion interactions. As an ionic material, we will look at sodium fluoride. Here, both the $Na^+$ cation and the $F^-$ anion are isoelectronic, and both ions are isoelectronic with an atom of neon. However, the interatomic interactions could not be more different.

When sodium and fluorine interact, they do so vigorously. There is a transfer of an electron from the sodium atom to the fluorine atom; thereby giving the two product ions a filled electron shell. The neutral sodium atom has a radius of about 1.54 Å, and when it has lost its outermost electron, the $Na^+$ has a radius of 1.16 Å. Conversely, the F atom has a radius of about 0.71 Å, and upon completing the $2p$ shell by accepting an electron the radius becomes 1.19 Å.

When these positively-charged and negatively-charged ions interact and coalesce, they form a crystal whose structure is seen in figure 5.1(a). This crystal has a classic ionic structure; that is, it is formed from ions held in place by electrostatic forces, rather than covalent bonds, or non-bonded interactions; sodium fluoride melts at 993 °C. As with neon, the structure is again cubic and a body-centred form with $a = b = c = 4.62$ Å; neon which is uncharged and isoelectronic with $Na^+$ and $F^-$ forms, as we saw earlier, a cubic close-packed structure with $a = b = c = 4.429$ Å, which melts at −248.59 °C. The crystals of these two very different materials are both cubic and of similar dimensions, but they are held together by the strongest (in the case of NaF) and the weakest forces (for neon) seen in chemical structures.

Sodium fluoride is a typical ionic crystal, where the structure is maintained by the powerful electrostatic interactions of the cations and the anions. The cations and anions are ordered so as to maximize the attractive electrostatic interaction between oppositely charged ions, and to minimize the repulsive interactions between ions of like charge. Coulomb's law, or Coulomb's inverse-square law describes the force arising through the interaction of electrically-charged particles. The law was first published in 1784 by French physicist Charles Augustin de Coulomb, and was a

(a)

(b)

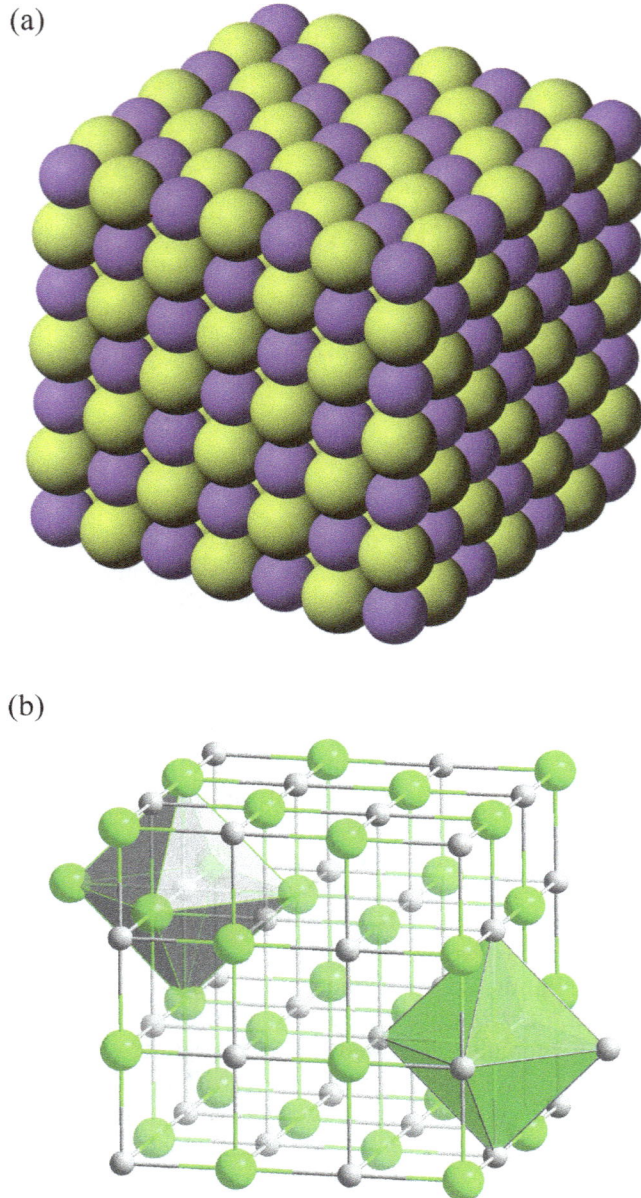

**Figure 5.1.** Part (a): the crystal structure of sodium fluoride. The sodium ions are in blue and the fluorine anions are in green. (Image from https://en.wikipedia.org/wiki/Sodium_fluoride#/media/File:Sodium-fluoride-3D-ionic. png). The crystal belongs to the cubic system. This is the halite or rock-salt structure (exemplified by sodium chloride, which is displayed in part (b)), in which each of the two ion types forms a separate face-centred cubic lattice, with the two lattices interpenetrating so as to form a 3-dimensional checkerboard pattern. Alternatively, one could view this structure as a face-centred cubic structure with secondary atoms in its octahedral holes. Examples of compounds with this structure include sodium chloride itself, along with almost all other alkali halides, and many divalent metal oxides, sulphides, selenides, and tellurides. Generally, this structure is more likely to be formed if the cation is somewhat smaller than the anion (a cation/anion radius ratio of 0.414 to 0.732).

major step in the development of the theory of electromagnetism. It is analogous to Isaac Newton's inverse-square law of universal gravitation. In its scalar form, the law is:

$$F = k_e(q_1\, q_2/r^2),$$

where $k_e$ is Coulomb's constant (it is a measure of how the electric potential propagates through space; $k_e = 1/(4\pi\varepsilon_0) = 8.99 \times 10^9$ N m$^2$ C$^{-2}$), $q_1$ and $q_2$ are the signed magnitudes of the charges, and $r$ is the distance between the charges. The force of interaction between the charges is attractive if the charges have opposite signs (that is, $F$ is negative) and repulsive if like-signed ($F$ is positive).

This electrostatic interaction is so strong that whenever an ensemble of sodium ions and fluoride ions organize to form a crystal, it dominates over other interactions. If one ion is much bigger that the other ion, however, there will be a significant additional force that arises between the interacting ions. This additional force is generated from the polarization effects of the electric field generated by the smallest, and most highly charged ion; that is, Fajan's rules.

From Coulomb's law, the electrostatic attraction between a single cation (of charge $e$) and a single anion (of charge $-e$), separated by a distance $r$, may be written as

$$U(r) = -e^2/r. \qquad (5.1)$$

As the distance between the ions increases, the electrostatic potential falls. If the ions are forced together, the electrostatic potential will rise, but only until a point where the ions become sufficiently close for their electrons to sense each other. At this point, there is a strong repulsion between the ions, which quickly overcomes the Coulomb attractive potential (see figure 2.6). The manner in which this short-range repulsive interaction between the ions varies is usually written as

$$U(r) = A/r^n, \qquad (5.2)$$

where $n$ is an integer, of order, 9 to 12, and $A$ is a constant. Consequently, the net potential energy for the pair of ions is the sum of the attractive term and the repulsive term; that is,

$$U(r) = (-e^2/r) + (A/r^n). \qquad (5.3)$$

These are the equations that a cloud of gaseous sodium cations and fluoride anions would follow when they self-assemble to form a crystal structure; the ions will move together along energy surfaces similar to the curve displayed in figure 2.6. But to calculate the electrostatic interaction energy in the entire crystal, as opposed to a single pair of ions, one must consider a single ion and consider its interaction with all the ions around it. For the attractive part of the inter-ionic potential, this calculation requires summing the interactions over many ions out to large distances from the central ion. As far as the short-range repulsive interactions are concerned, it usually suffices merely to consider nearest neighbour interactions, as the repulsive energy falls off so quickly with distance (varying as $1/r^n$ with large values for $n$).

Within the crystal, the distance $r_i$ from the ion under consideration to any other ion i in a surrounding shell can be expressed in terms of the nearest neighbour distance $r_0$ in the crystal structure: $r_i = p_i r_0$, where $p_i$ is a dimensionless constant for the crystal structure under consideration. The total attractive electrostatic energy, $E_T$ is then

$$E_T = \sum_i (\pm e^2/r_i) = -\alpha e^2/r_0, \qquad (5.4)$$

where the constant $\alpha$ is termed the Madelung constant, and this energy is termed the Madelung energy. If the central ion under consideration is a cation, the terms in the summation are positive for anions i, and negative for cations i.

The Madelung constant is used in determining the electrostatic potential of a single ion in a crystal by approximating the ions by point charges. It is named after the German physicist Erwin Madelung (1881–1972). Because the anions and cations in an ionic solid are attracting each other by virtue of their opposing charges, separating the ions requires equivalent energy. This energy must be given to the system in order to break the anion–cation bonds (what could be termed ionic bonds). The energy required to break these bonds for one mole of an ionic solid under standard conditions is the lattice energy of the crystal.

The six fluoride anions at $r_0$ from the sodium cation ($p_1 = 6$) contribute 6 to the above summation, equation (5.4). Then around this first shell of ions, there are 12 $Na^+$ ions at $p_2 = 2^{1/2}$, giving a contribution of $-12/2^{1/2}$ to the summation. Next, there are $F^-$ ions at $p_3 = 3^{1/2}$, which give a contribution of $8/3^{1/2}$, and so on. It must be stated that this is a very slowly converging series, and so many terms need to be considered. The repulsive term for the inter-ionic potential is a short-range interaction when compared with the Coulombic forces, and it is usually only necessary to consider the nearest six neighbouring ions.

Crystal Engineering
How molecules build solids
**Jeffrey H Williams**

# Chapter 6

# Materials with mixed bonding

As mentioned earlier, there is no such thing as pure ionic bonding. Even in alkali halides, ionic bonds have a covalent character, and hence one should talk about mixed bonding in inorganic materials.

## 6.1 Ruby

Although prized as red gemstones, ruby is a species of white corundum, or $\alpha$-$Al_2O_3$ coloured red by a small amount of chromium dispersed throughout the crystal. Indeed, sapphires are a blue variety of the same corundum, where the colour is generated by iron and/or titanium dispersed throughout the crystal.

Corundum crystallized in the trigonal system, with the structure seen in figure 6.1. This structure belongs to the trigonal-hexagonal scalenohedral class of the trigonal system, and has the following unit cell dimensions: $a = b = 4.764$ Å and $c = 13.009$ Å. The rhombohedral angle is 55°17'; the unit cell of the crystal contains two $Al_2O_3$ units, so the occupancy ($Z$) is 2. Although one speaks of two units of $Al_2O_3$ per unit cell, there are in fact no discrete *molecules* of $Al_2O_3$ to be seen in the crystal; this formula is the average of the chemical composition of the subunit of the solid. The structure displayed in figure 6.1 can be seen to be constructed of edge-shared octahedral, containing an aluminium ion and a fraction of the six surrounding $O^{2-}$ ions. The individuality of the $Al_2O_3$ units has been lost in the ionic nature of the crystal; but as we saw earlier, there is no such thing as a pure ionic bond, and this crystal has inter-ionic links that are partially ionic and partially covalent. We can better imagine this structure as an approximately hexagonally close-packed array of oxide ions ($O^{2-}$) with $Al^{3+}$ ions in some of the octahedral interstitial sites. There are sufficient octahedral sites between each layer of oxide ions to accommodate one $Al^{3+}$ ion for each $O^{2-}$ ion. Thus, in corundum only two thirds of these sites can be filled.

The octahedral voids being occupied by the $Al^{3+}$ ions are far from regular (this can be seen best in figure 6.1(a)). Three oxide ions lie in a plane 1.37 Å from the $Al^{3+}$ ion, while the other three oxide ions lie in a plane that is only 0.8 Å from the $Al^{3+}$

(a)

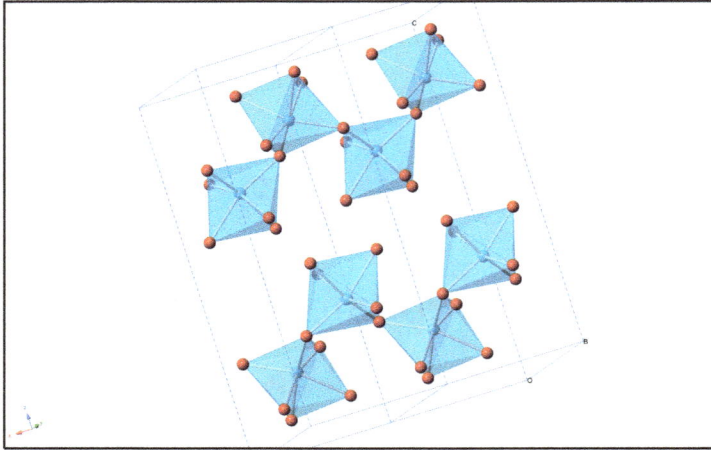

(b)

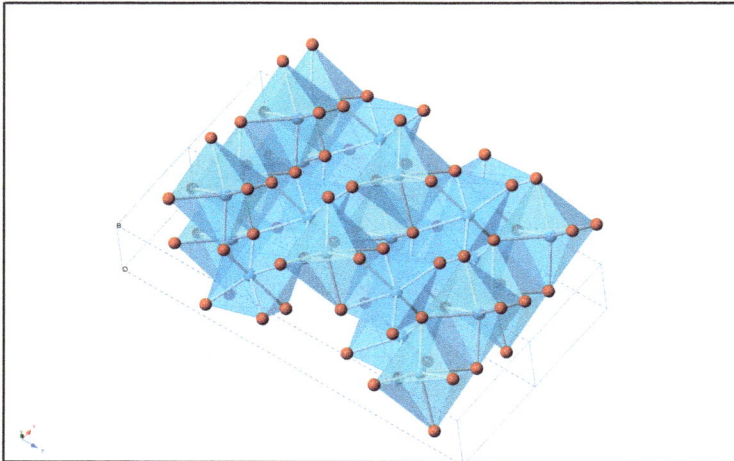

**Figure 6.1.** Two views of the structure of corundum; the blue spheres in the transparent blue octahedra are aluminum (global) ions ($Al^{3+}$) and the red spheres, which make up the octahedra are oxygen ions (oxide ions, $O^{2-}$). The differences between these two images are the number of octahedra displayed, and the orientation.

ion. This gives Al–O distances (given that this solid is mostly built of ionic bonds, it is best to speak of distances in the solid, rather than bond lengths) of 1.98 Å and 1.84 Å, respectively.

The colour is imparted to ruby by chromium ions dispersed throughout the structure; pink rubies have about 0.04% of their $Al^{3+}$ exchanged by $Cr^{3+}$, and deep red rubies have 0.5% of their aluminium ions replaced by $Cr^{3+}$. Chromium ions are able to enter the structure as they have a similar ionic radius (0.69 Å in $Cr^{3+}$ and 0.57 Å in $Al^{3+}$) and the same electrical charge as aluminium ions.

As mentioned above, the $Cr^{3+}$ ions replace some $Al^{3+}$ ions in the crystal, and so these chromium ions are to be found in these distorted octahedral sites between layers of oxide ions. In such positions, the energy levels of the $Cr^{3+}$ ions are perturbed by the intense electric fields generated by the asymmetric environment of oxide ions; the electric field arising from the layer of oxide ions closest to the chromium ions will not be cancelled by the electric field arising from the layer of oxide ions on the other side of the chromium ion, as that layer is further away (an electric field is charge/distance). Consequently, the energy levels of the $Cr^{3+}$ ion will be distorted by this local crystal field. This local field is so strong that it generates new spectroscopic features in the chromium ions that are not seen in isolated $Cr^{3+}$ ions.

## 6.2 The crystal field in ruby

The $Cr^{3+}$ ion has an electronic configuration of $1s^2 2s^2 2p^6 3s^2 3p^6 3d^3$, and consequently there are three (3d) electrons outside a closed-shell, argon-type structure. Each of these electrons move in a field generated by the presence of the other electrons in the ion, and in the field of the positively-charged nucleus of the $Cr^{3+}$ ion. So, the total energy of an electron (of kinetic energy $p^2/2m$, where $m$ is its mass and $p$ its momentum) in the $Cr^{3+}$ ion, $E_T$, will include the repulsive interaction of the other electrons (proportional to $e^2/r$, where $e$ is the electronic charge and $r$ is its distance from the electron of interest) and the attractive interaction of the nucleus (proportional to $-Ze^2/r$, where Z is the nuclear charge and $r$ is the distance to the nucleus from our electron). Thus

$$E_T = (p^2/2m) - (Ze^2/r) + \sum(e^2/r_i), \qquad (6.1)$$

where $r_i$ is the distance between our electron and the other electrons.

In addition to these large energy contributions, we also have to consider the contribution of the spin of the electron under consideration; this is a weaker interaction but still a non-negligible contribution; termed the spin–orbit interaction. The electron has an intrinsic spin, which is what determines how they couple together on an individual energy level, and which can be thought of as a small permanent magnet. As the electron moves in its orbit around the nucleus, it acts like a current in a coil of wire, and the bigger the radius of the electron's orbit (3d electrons have larger radii than 2p electrons) the bigger is the magnetic field generated, which gives the electron an orbital magnetic moment as well as its spin magnetic moment. The interaction between these two magnetic moments is the spin–orbit interaction, and it adds a term to equation (6.1), $E_{spin-orbit} = \lambda$ **l.s**, where **l** is a quantum number specifying orbital angular momentum, **s** is the spin quantum number, and $\lambda$ is the spin–orbit interaction constant[1].

---

[1] **l** and **s** are written as vectors to indicate that the interaction depends upon the orientation of the two magnetic moments. A dot product of two vectors is a scalar representation, $l.s \cos\theta$, where $\theta$ is the angle between $l$ and $s$; so $s.\cos\theta$ is the component of s in the direction of l.

Consider the three $3d^3$ electrons of the $Cr^{3+}$ ion. These electrons can be classified according to the arrangements of their spins, and the total spin of the system, $S$. If all three spins are parallel: ↑↑↑, then $S = 3/2$ and $2s + 1 = 4$, and we speak of *quartets*, but if only two of the spins are parallel: ↑↓↑, the $S = ½$ and $2s + 1 = 2$, and we speak of *doublets*. The term, $2s + 1$ is the multiplicity of the arrangement, or the number of ways that the total spin can be aligned in an applied field. The component of $S$ in the field direction is specified by the quantum number $M_S$; for example, for $S = 3/2$, $M_S$ can be 3/2, 1/2, −1/2, −3/2 (a quartet); for $S = 1/2$, $M_S$ is either 1/2 or −1/2, a doublet.

The lowest energy state, or ground state for an isolated $Cr^{3+}$ ion is a quartet denoted by 4F. In accord with Hund's rule (named for the German physicist Friedrich Hund, 1896–1997), the state of highest multiplicity will be the ground state, because when the electrons have their spins parallel, their spin quantum numbers are identical and they are forced by the Pauli exclusion principle to avoid each other. Consequently, repulsion between the electrons, which would raise the potential energy of the system is kept to a minimum. The energy level scheme is shown in figure 6.2.

If the $Cr^{3+}$ ion is placed in a distorted octahedral site in the corundum crystal, it will be subject to an intense local or crystal field, $V_{cf}$, and this field will modify the

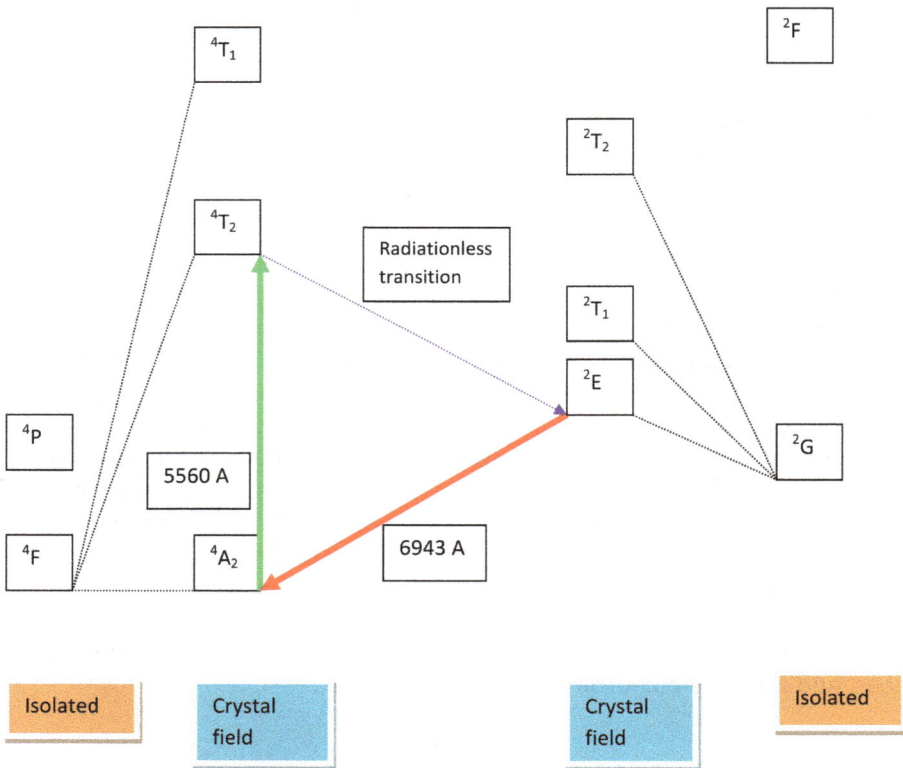

**Figure 6.2.** The energy levels of a $Cr^{3+}$ ion in the isolated ion and in the ion, as modified by the perturbation of the crystal field generated by the oxide ions. See text for details.

coupling of the spins in the chromium ion. If the local field interacts only slightly with the $3d^3$ spins, then the energy levels of the ion in the crystal will be much the same as the energy level of the free ion; that is, the visible spectrum will be largely unchanged. If, however, the interaction energy, $E_{interaction}$, of the local field with the $3d^3$ electrons is of the order of the spin–orbit coupling, then the energy levels of the $Cr^{3+}$ ion in the crystal will be very different from those of the free ion and the spectrum will change significantly.

For the chromium ions in ruby, the crystal field energy of interaction is larger than the spin–orbit interaction, but considerably smaller that the electrostatic interaction between the electrons in the chromium ion; that is, $e^2/r_i > E_{interaction} > \lambda \, \mathbf{l} \cdot \mathbf{s}^2$.

The effect of the local field generated by the oxide ions on the lowest quartet level $^4F$ and the lowest doublet $^2P$ level of an isolated $Cr^{3+}$ ion are shown in figure 6.2. The transition between the ground state $^4A_2$ and the excited $^4T_2$ state corresponds to a wavelength of 5560 Å, in the green part of the spectrum. As the $Cr^{3+}$ ions, in the presence of the crystal field, absorb light in the green part of the spectrum, the crystal appears red; the colour of ruby is thus explained by an optical absorption induced by a local crystal field. This transition is allowed as the two states have the same number of unpaired spins; and hence there is no absorption band corresponding to the transition from the $^4A$ state to the $^2E$ state.

This local field-induced absorption in the green has another consequence; a ruby crystal was one of the first solids to be seen to display laser action. Remember that the word laser is an acronym: Light Amplification by Stimulated Emission of Radiation. And the explanation of laser action takes us back to Albert Einstein's quantum mechanical explanation of the absorption of radiation by matter.

The local field of the oxide ions induces an absorption ($^4A_2 \rightarrow {}^4T_2$) at 5560 Å. As a beam of light at a wavelength of 5560 Å passes through a ruby, it is absorbed and $Cr^{3+}$ ions are excited into the $^4T_2$ level, leading to an enhanced concentration of ions in this excited state. Most of this absorbed ration is not re-emitted at the same wavelength; there is another transition, which occurs at a more rapid rate than the emission process. This is the radiationless transition to the $^2E$ state, see figure 6.2. This transition is said to be phonon assisted, because the transfer of the energy comes about through the thermally-driven vibrations of the crystal lattice[3].

The $^2E$ state is said to be metastable with respect to the ground state of the system, $^4A_2$; the spontaneous emission coefficient is only about $10^2 \ s^{-1}$, and so if enough power is pumped into $^4A_2 \rightarrow {}^4T_2$ transition, a population inversion may be created in the long-lived $^2E$ state; that is, the population of exited $Cr^{3+}$ ions in this state is

---

[2] If $E_{interaction} \gg e^2/r_i$ we would have the situation seen in systems where ligands or groups are bound covalently to the centre metal ion; for example, in the ferricyanide ion. Here, the Fe–CN covalent bonds dramatically change the electronic configuration of the $d$ electrons in the $Fe^{3+}$ ion located at the centre of the octahedral [Fe(CN)$_6$]$^{3-}$. If $E_{interaction} \ll e^2/r_i$, it is likely that the electrons of the central ion are so well shielded that the crystal field interaction is small even compared to the spin–orbit interaction. This is seen in crystals that contain rare earth ions, in which the $4f$ electrons have nearly the same energy in the crystal as they do when isolated.
[3] The rate of the transition $^4T_2 \rightarrow {}^4A_2$ is, of order, $3 \times 10^5 s^{-1}$, but the transition $^4T_2 \rightarrow {}^2E$ has a rate, or order, $10^7 s^{-1}$.

higher than would be expected from the Boltzmann equation. If such a non-equilibrium population of excited ions is exposed to some red light at 6943 Å, the induced emission will rapidly deplete the excited population. There will be an intense burst of radiation at 6943 Å as the ions in the $^2$E state return to the ground state by a process of stimulated emission. Light amplification would then have been accomplished, and the ruby crystal would have *lased*.

## 6.3 Calcite

This is a mineral composed of spherical positively-charged calcium ions ($Ca^{2+}$), and flat, triangular carbonate ions ($CO_3^{2-}$), which are negatively-charged. In sodium fluoride, we had an exchange of one electron between the sodium and the fluorine atoms, but in calcite two electrons have been exchanged (the two charges are distributed, or delocalized over the large flat carbonate anion). This material is well-known as it occurs naturally as marble, limestone, chalk and as the beautiful crystals of calcite, known as Iceland spar. The crystals of calcite are formed from columns and rows of $Ca^{2+}$ and $CO_3^{2-}$ ions, see figure 6.3. Calcite is white or transparent, though shades of grey, red, orange, yellow, green, blue, violet, brown, or even black can occur when the mineral is charged with impurities. It was named as a mineral by Gaius Plinius Secundus (Pliny the Elder) in 79 AD from Calx, the Latin for lime. Calcite belongs to the trigonal system and to the calcite group; it is the defining member of that group. The unit cell of calcite is characterized by the dimensions, $a = b = 4.9896(2)$ Å and $c = 17.0610(11)$ Å; and the occupancy number of the unit cell, $Z = 6$.

Calcite is an ionic material, the structure of which resembles that of NaF; the place of the spherical $F^-$ ions being taken by planar $(CO_3)^{2-}$. These large planar groups are arranged in parallel position in the crystal, with their 3-fold symmetry

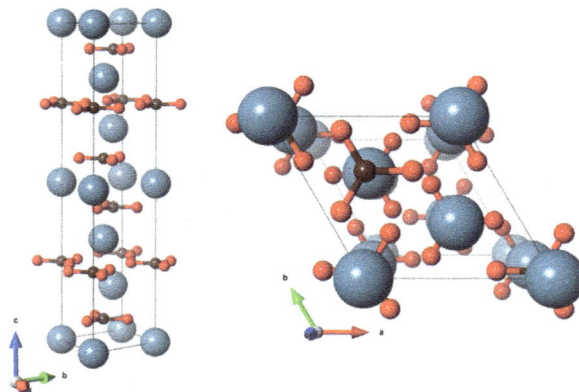

**Figure 6.3.** Two views of the structure of calcite, which is made up of three atoms, calcium (blue), carbon (brown) and oxygen (red). The structure consists of calcium ($Ca^{2+}$) cations sandwiched between flat layers of carbonate ($CO_3^{2-}$) anions. Calcite belongs to the trigonal system and to the hexagonal scalenohedral crystal class. See http://www.chemtube3d.com/solidstate/SS-CaCO3.htm for an interactive animated structure of calcite and closely related minerals.

axes oriented along $[0001]^4$. This arrangement leads to the strong birefringence of calcite-group minerals. This natural birefringence causes objects viewed through a clear piece of calcite to appear doubled.

Calcite is often said to possess a modified cubic structure. Calcite is not cubic, and it is the presence of the anisotropically polarizable $CO_3^{2-}$ anion that distorts the cubic structure. The crystal structure of calcite is illustrated in figure 6.3 and consists of a rather complex, but highly ordered 3-dimensional array of $Ca^{2+}$ and $CO_3^{2-}$ ions. Again, the overall structure is determined essentially by the electrostatic interaction of the positive and negative electrical charges on the two ions, which here is a larger interaction than in sodium fluoride as two electrons have been transferred. In calcite, however, the final structure is not simply generated by a balancing of the electro-static repulsion and electrostatic attraction between two electrical charges. The $CO_3^{2-}$ anion has an intrinsic structure (it is a triangular, planar arrangement of four atoms, over which its double electric charge is delocalized) and this anisotropic anion, which is large and full of electrons that are subject to the polarizing effects of the electric field generated by the $Ca^{2+}$ ions, will make a significant contribution to the overall electromagnetic forces that generate the structure seen in figure 6.3 (see equation (2.3) for the manner in which a molecule can become polarized by an applied electric field).

As a consequence of the presence of this anisotropic polyatomic unit, calcite has an anisotropic crystalline structure that interacts with light in a totally different manner than an isotropic crystal such as sodium fluoride. Indeed, crystals are classified as being either isotropic or anisotropic depending upon how light behaves on propagating through the crystal, which tells us whether or not the crystal's crystallographic axes are equivalent or not. All isotropic crystals (such as sodium fluoride) have equivalent axes that interact with light in a similar manner, regardless of the crystal's orientation with respect to incident light waves; that is, the distribution of polarizable material (electrons) in the crystal is uniform, or isotropic.

We see clearly in figure 6.3, however, that the anisotropic $CO_3^{2-}$ ions are all aligned in the crystal. This generates a coherence which is manifest as a new bulk property of the solid, birefringence. The molecular planes of the planar $CO_3^{2-}$ anions are parallel to each other and perpendicular to the main crystal axis. Thus, the anisotropic nature of the distribution of electrons in the $CO_3^{2-}$ anion is imprinted on the macroscopic calcite crystal.

In optics, it is the refractive index ($n$) of a substance that is the basis of investigating how light is transmitted by that material. This index is a dimensionless number, being the ratio of two identical quantities. It is the ratio of the speed of light in a vacuum ($c$) and the speed of light in the material of interest ($v$), $n = c/v$. For example, the refractive index of water in the visible part of the spectrum is 1.333, meaning that light travels 1/1.333 times slower in water than it does in a vacuum. It is assumed that the vacuum is the medium which offers the least resistance to the passage of light (the least dense of all possible media), but as we know, the quantum mechanical view of the vacuum tells us that it is full of quantum fluctuation.

---

[4] A fourth index is required in trigonal/hexagonal systems to assist with the specification of directions within the crystal; *hkl* and *i*, where $i = -h-k$.

Birefringence, or double refraction of light arises from differences in refractive index within a crystal. In calcite there is a difference in the refractive index of the crystal (or in terms of the molecular components of the crystal, the polarizability, which is the quantum mechanical explanation of how light propagates through matter) due to the planar $CO_3^{2-}$ anions. The directions parallel and perpendicular to the main rotational axis of the anion coincide with an axis of the crystal. When light enters such a non-equivalent axis system, it is refracted into two rays, each polarized with the vibration directions oriented at right angles to one another and travelling at different velocities. This phenomenon is termed double refraction and is exhibited to a greater or lesser degree in all anisotropic crystals. There are, in effect, two routes for light to pass through a birefringent crystal that arise because of the anisotropic polarizability of one of the component groups of the crystal; the anisotropic polarizability of the $CO_3^{2-}$ anions.

It is in calcite that we see the most dramatic example of a natural birefringence (at a wavelength of 5900 Å, calcite has ordinary and extraordinary refractive indices of 1.658 and 1.486, respectively.). The calcite crystal produces two images when it is placed over an object and then viewed with reflected light passing through the crystal. One of the images appears as would normally be expected when observing an object through clear glass or an isotropic crystal, while the other image appears slightly displaced, due to the nature of doubly refracted light. Thus, when we move from sodium chloride to calcium carbonate, the forces holding the structures together are both electromagnetic in origin, but are more complex in the case of calcite due to the presence of the anisotropically polarizable $CO_3^{2-}$ ions. However, this additional complexity generates a new property of the macroscopic crystal, birefringence. We see here how the organization of electrons in a planar structure of four atoms can generate a bulk optical effect. Here, there is order as well as function arising from the interacting atoms, and the electromagnetic forces they generate.

## 6.4 Beryllium fluoride

Beryllium forms binary compounds with many non-metals. $BeF_2$ has a silica-like structure with corner-shared $BeF_4$ tetrahedra. $BeCl_2$ and $BeBr_2$ have chain structures with edge-shared tetrahedra. All beryllium halides have a linear monomeric molecular structure in the gas-phase; that is, above its boiling point of 1169 °C, F–Be–F is linear, however, when these monomers interact on cooling they generate a structure more reminiscent of the condensation of an ionic species than a structure form from apparently covalent monomers. The polymeric structure of the solid, where there are no individual F–Be–F molecules, arises through the formation of dative bonds (a form of covalent bonding) between the lone pairs of electrons on the fluorine atoms and the beryllium atoms in neighbouring $BeF_2$ molecules.

Beryllium forms covalent compounds through hybridization of its atomic orbitals. The ground state electronic configuration of Be is $1s^2\,2s^2$ and it may be thought of as undergoing $sp$-hybridization to form $BeF_2$. The electronic configuration of fluorine is $1s^2\,2s^2\,2p_x^2\,2p_y^2\,2p_z^1$ which has one half-filled $p$-orbital. The hybridized beryllium has two half-filled $sp$-orbitals, and the half-filled $p$-orbital of fluorine overlaps with one of the *sp*-hybrid orbitals of Be to form a single $\sigma$-bond between Be and F. The other F atom also forms a single $\sigma$-bond with Be. The two bonds are opposite to each other, and so the molecule is linear. Each of these bonds will have a strong bond dipole moment as the two atoms have very different electronegativities. However, these two bond dipole moments will oppose each other and the vector sum will be zero. But this gives the linear $BeF_2$ molecular a large negative quadrupole moment like $CO_2$; that is, the partial charges may be written as $(\delta-)F-Be(2\delta+)-F(\delta-)$.

The structure of $BeF_2$ can be seen in figure 6.4, and it is evident that there are no discernible monomers. The monomers have adopted, at low temperatures, a structure that is neither purely ionic, nor purely covalent. In fact, there has been a polymerization of the monomers, and addition bonds have been formed between the fluorine atoms and the beryllium atoms. These addition bonds polymerize the covalent linear triatomic molecules, stable at high temperature, giving the extended crystalline structure that exists at ambient conditions. The structure of $BeF_2$ is polymeric giant structure, such are quartz, graphite and diamond, and is very different from that of ionic calcium fluoride, as seen in figure 6.5, which has a discrete unit cell that is repeatedly to generate the crystal architecture.

Beryllium fluoride crystallises in the trigonal system and has the following unit cell dimensions: $a = b = 4.739$ Å, and $c = 5.179$ Å. A material with a similar structure is the low-temperature form of quartz ($\alpha$-quartz), where the unit cell has dimensions $a = b = 4.9133$ Å, and $c = 5.4053$ Å and the crystal is built up of $SiO_4$ tetrahedra as opposed to $BeF_4^{2-}$ tetrahedral.

Beryllium difluoride, $BeF_2$, is different from the difluorides of closely related elements. In general, beryllium has a greater tendency to bond covalently than the other alkaline earths and its fluoride is partially covalent (although still more ionic than its other halides). Beryllium fluoride melts at 554 °C and boils at 1169 °C, while magnesium fluoride is more ionic melting at 1262 °C and boiling at 2260 °; calcium fluoride melts at 1418 °C and boils at 2533 °C (steel melts at a mere 1400–1500 °C). On descending this group of elements, for a constant dihalide, the gas-phase or isolated molecules change from being centrosymmetric linear ($BeF_2$, $MgF_2$) to bend, $C_{2v}$, ($CaF_2$). This change in molecular geometry in the isolated molecules is believed to occur due to an electronic curve crossing in the middle of the group. As the metal atoms become larger, the relative positions of the $2A_1$ and $B_2$ type molecular orbitals change, and these levels cross in the calcium atom. Whilst the difference in energy between the $2A_1$ and the $B_2$ states is large for both Be and Mg, and for Ba, for $CaF_2$ this energy difference is small and so the structure of calcium fluoride (in the isolated molecule) is subject to perturbations; perturbations of order of $k_BT$. Of course, in the solid state of an ionic material, it is not possible to talk about isolated molecules having a particular form, but the crossing of the electronic levels of the metal atoms needs to be considered in any analysis of the structure of the unit cell.

Beryllium fluoride has many similarities to $SiO_2$ (quartz) a mostly covalently bonded network solid. $BeF_2$ has tetrahedrally coordinated metal atoms and readily forms disordered glasses. When crystalline, beryllium fluoride has the same room temperature crystal structure as $\alpha$-quartz (both $BeF_2$ and $\alpha$-quartz crystallize in the

(a)

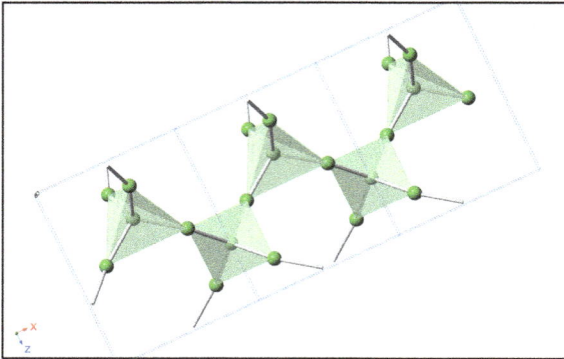

(b)

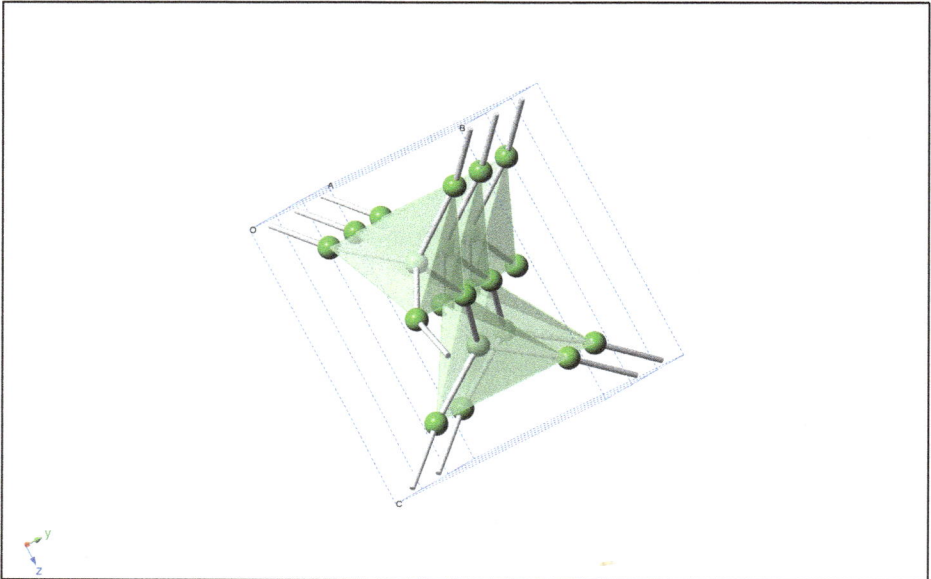

**Figure 6.4.** Two views of the structure of beryllium fluoride; the fluoride ions (the darker green) are actually larger than the beryllium ions, which are pale green and located inside tetrahedra of fluoride ions.

trigonal system). Beryllium difluoride is soluble in water, unlike the other alkaline earth difluorides, which are naturally occurring minerals. (Although they are strongly ionic, they do not dissolve because of the especially strong lattice energy of the fluorite structure.) However, $BeF_2$ has much lower electrical conductivity when in solution or when molten than would be expected if it were fully ionic. It is a material that is a mixture of covalent and ionic properties, representing a transitional state between covalent and ionic bonding.

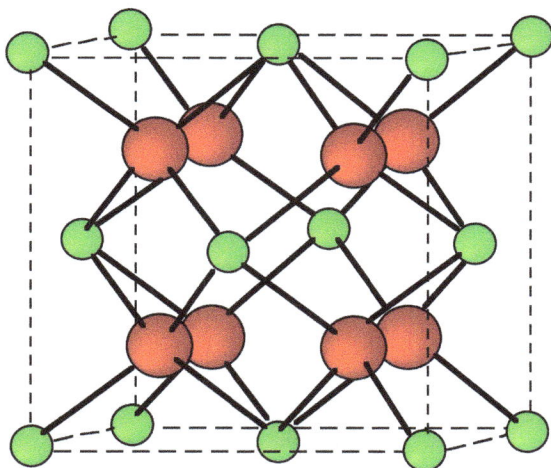

**Figure 6.5.** Crystal structure of calcium fluoride or fluorite (Image from http://archive.cnx.org/contents/ ae596b60-b3d5-44e2-9f5f-a66aae8206e4@3/calcium-the-archetypal-alkaline-earth-metal). The fluoride ions are green and shown occupying the corners of the cube, but it could equally well be drawn with the calcium ions (red) occupying the corners of the cube. The crystal belongs to the cubic system and is characterized by a unit cell length of $a$ = 5.462 Å. See http://www.chemtube3d.com/solidstate/_fluorite(final).htm for interactive animated structure.

What we are seeing in $BeF_2$ is a loss of individuality in the bulk material; usually ionic crystals represent systems that cannot exist under normal conditions as monomers, even as $Na^+Cl^-$ at high temperatures. In covalent solid nitrogen, however, or solid benzene one can see the individual molecules, they are weakly bound together, but the monomeric units from which the giant extended structure is built can be clearly discerned. But in $BeF_2$ there is a change-over between these two regimens.

## 6.5 Lithium niobate

This is a colourless crystalline solid of chemical formula, $LiNbO_3$. It melts at the high temperature of 1257 °C. The bonding in this material is therefore a mixture of ionic and covalent. It is a mixture of atomic ions and ionic polyatomic units, containing niobium ions bonded to an octahedral arrangement of oxygen ($O^{2-}$) ions; see figure 6.6.

The crystal belongs to the trigonal system, and is characterized by the following unit cell dimensions: $a = b$ = 5.1505 Å and $c$ = 13.8649 Å. This crystal symmetry means that it lacks an inversion symmetry and so is able to display a range of properties that arise from the anisotropic distribution of polarizable material in the crystal; for example, lithium niobate displays ferroelectricity, the Pockels effect (a birefringence induced in a material by application of an applied electric field), the piezoelectric effect, photoelasticity and nonlinear optical polarizability. Lithium niobate has a negative uniaxial birefringence, which depends slightly on the

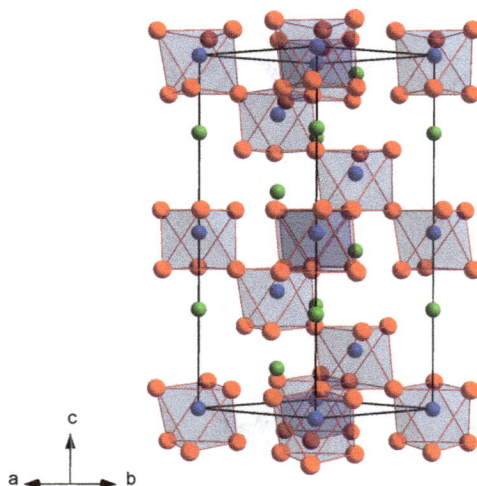

**Figure 6.6.** The structure of lithium niobate. The lithium atoms are green, the niobium atoms are blue and the oxygen atoms are red. See http://www.chemtube3d.com/gallery/structurepages/NbO3-poly.html for an interactive animation of the niobate anion.

stoichiometry of the crystal and on temperature. It is transparent for wavelengths between 3500 and 52 000 Å. Consequently, lithium niobate is a material of central importance in the electronics and optoelectronics industries.

## 6.6 Piezoelectricity

Lithium niobate is a crystal with several remarkable properties, not the least of which is its ability to generate an electrical potential when squeezed; that is, a piezoelectric effect.

Piezoelectricity is the accumulation of electric charge in certain solids in response to an applied mechanical stress. It is thus a property of crystals whose internal structure, or unit cells are susceptible to applied pressure. The discovery of piezoelectric effect occurred in 1880, and was made by French physicist and Nobel Laureate Pierre Curie (who also developed a theory for ferromagetism) and his brother Jacques. At that time, quartz and Rochelle salt (potassium sodium tartrate tetrahydrate) were also found to display a piezoelectric effect. One year later, the converse piezoelectric effect was predicted theoretically by the French physicist, inventor of colour photography and Nobel Laureate, Gabriel Lippmann.

The piezoelectric effect is the linear electromechanical interaction between the mechanical and the electrical state of a crystal without inversion symmetry. The piezoelectric effect is reversible, in that materials exhibiting the direct piezoelectric effect (the internal generation of electrical charge resulting from an applied mechanical force) also exhibit the reverse piezoelectric effect. For example, lead zirconate titanate crystals, of general formula $Pb[Zr_xTi_{1-x}]O_3$ ($0 \leqslant x \leqslant 1$), will generate measurable piezoelectricity when their static structure is deformed by about

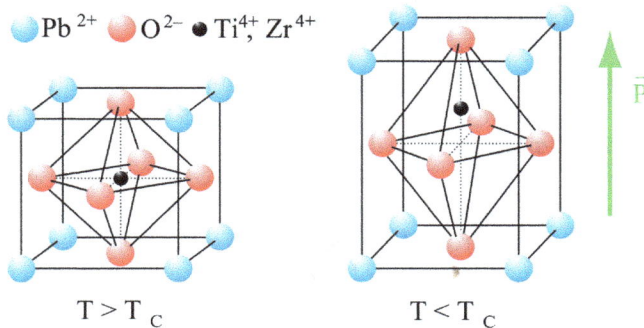

**Figure 6.7.** An explanation of piezoelectricity. A schematic representation of a unit cell showing a perovskite crystal structure (here, lead zirconate titanate), which demonstrates the origin of piezoelectricity. On the left is the cubic form of the crystal and on the right is the tetragonal form; the material is tetragonal below $T_c$ and cubic above $T_c$. In the tetragonal phase, the spontaneous polarization $P$ is indicated. See http://www.chemtube3d.com/solidstate/_perovskite(final).htm for an animated interactive structure for the archetype structure for this behaviour, perovskite.

0.1%. Conversely, those same crystals will change about 0.1% of their static dimension when an electric field is applied externally[5].

The nature of the piezoelectric effect is related to the occurrence of an electric dipole moment in the crystal. In mixed ionic-covalent materials, such a dipole may arise either through an asymmetric arrangement of ions on crystal lattice sites, or may directly be carried by molecular groups (as in the piezoelectric effect of sucrose, which forms monoclinic crystals, of point group P2(1) with unit cell $a = 10.89$ Å, $b = 8.69$ Å and $c = 7.77$ and $\beta = 103°$). The dipole density or polarization (that is, the electrical charge per unit volume, or $Cm^{-3}$) may be calculated for crystals by summing up the dipole moments per volume of the crystallographic unit cell. As every dipole is a vector, the dipole density $P$ is a vector field. An ensemble of electric dipoles (and magnetic dipoles in ferromagnetism) tend to be aligned into regions or domains via an electrostatic dipole–dipole coupling.

Of importance for the piezoelectric effect is the change of polarization when applying a mechanical stress. This might either be caused by a pressure-induced reconfiguration of the dipole-inducing structural elements in the crystal structure, or by re-orientation of molecular dipole moments under the influence of the external stress. Piezoelectricity may then be manifest in a variation of the polarization strength, its direction or both, with the details depending upon: (1) the orientation of $P$ within the crystal, (2) crystal symmetry, and (3) the applied mechanical stress. The change in $P$ appears as a variation of surface charge density upon the crystal faces; that is, as a variation of the electric field extending from the faces caused by a change in polarization of the bulk.

The mechanism responsible for the piezoelectric response of certain materials can be clarified by looking at the structure given in figure 6.7. Piezoelectric materials are

---

[5] For example, a 1 cm$^3$ cube of quartz subject to a force of 2 kN (a force equivalent to about 200 kg) can produce a potential of 12 500 volts.

not electrically symmetric, and there is always a potential energy when opposite electrical charges interact, as one moves away from electrical neutrality. In figure 6.7, we see a crystal in an un-deformed (electrically neutral) state, and the same material when deformed; in the latter situation we see that the distortion of the lattice has led to an asymmetric distribution of the positive and negative charges in the octahedron of $O^{2-}$ ions; under the influence of the applied stress, the octahedrally coordinated titanium and zirconium ions have been displaced away from the centre of the octahedron of $O^{2-}$ ions. This generates an electric dipole in the unit cells, and hence a macroscopic polarization of the crystal, when all those tiny dipoles in each unit cell line up coherently.

The piezoelectric effect describes the relationship between stress, strain and the applied or produced electric field (or charge displacement) that is exhibited in some materials. The piezoelectric effect in crystals can be described by a pair of tensor equations for the strain and the electric field, often referred to as the direct and inverse relations.

## Further reading

For details about the application of symmetry to the investigation of crystal structures and the physical properties of crystals, please see *Crystal Symmetry and Physical Properties*, by Suri Bhagavantam, Academic Press, 1966. And the essentials are to be found in: S Bhagavantam and D Suryanarayama 1949 *Acta Crystalogr.* **2** 21; and H A Jahn 1949 *Acta Crystalogr.* **2** 30.

# Chapter 7

## Covalent solids

The majority of the systems we will look at in the remaining chapters fall into this category.

## 7.1 Nitrogen and carbon monoxide

Nitrogen is an inert, stable gas. Not only is gaseous nitrogen ($N_2$) a major component of our atmosphere, but it is also found in the atmospheres of neighbouring planets. Indeed, it is believed that solid nitrogen mixed with solid carbon monoxide and solid methane are to be found on the surface of distant, cold Pluto.

The electron configuration of the nitrogen molecule is, $- 2s\sigma^2\ 2s\sigma^{*2}\ 2p\sigma^2\ 2p\pi^4$. There are 6 net bonding electrons (so the molecular bond order is 3). Thus, the bond is stronger than that in $O_2$ and CO, and there are no unpaired electrons, so $N_2$ is diamagnetic. The bond between the two nitrogen atoms is short, at 1.098 Å; carbon monoxide has a bond length of 1.128 Å. The first non-vanishing electrical moment for nitrogen is the quadrupole moment, which has been measured to be $-5.25 \times 10^{-40}$ C m$^2$, this may be compared to the value for $CO_2$ of $-14.98 \times 10^{-40}$ C m$^2$, and we see that the phase of the value for nitrogen gives us the following picture of the molecule $(\delta-)N-(2\delta+)-N(\delta-)$; that is, the charge density is seen to be present at the ends of the molecule rather that off axis in the centre of the bond.

There are several known solid forms of molecular nitrogen. At ambient pressure there are two solid forms. β–$N_2$ is a hexagonal close-packed structure of the molecule centres of mass, which exists from 35.6 K up to 63.15 K at which point it melts. At 45 K, the unit cell has dimensions $a = b = 4.050$ Å and $c = 6.604$ Å. The crystal structure is susceptible to pressure, and at a pressure of 4125 atmospheres, and 49 K the unit cell has been compressed to $a = 3.861$ Å $c = 6.265$ Å.

Another phase, termed $\alpha$-$N_2$ exists below 35.6 K at ambient pressure and has a cubic structure (see figure 7.1 for the structure of this phase). At 21 K the unit cell

(a)

(b)

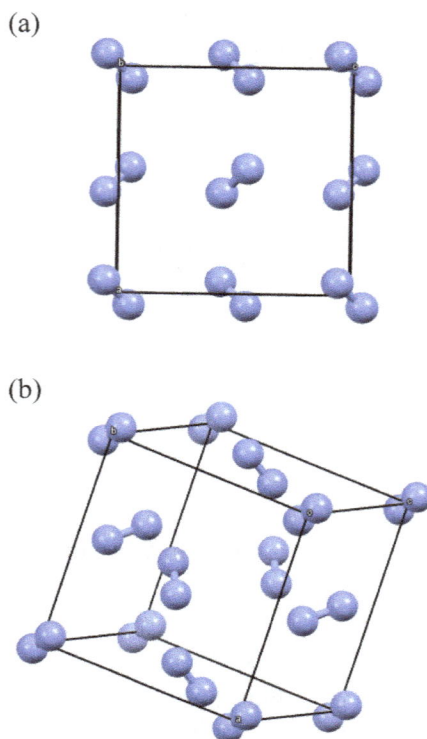

**Figure 7.1.** Two views of the structure of the lowest temperature phase of solid $N_2$, or dinitrogen ($\alpha$-nitrogen). The crystal belongs to the cubic system and is characterized by a unit cell of occupancy (Z) 4. Part (a) clearly demonstrates the classic *herringbone* structure displayed by structures formed primarily by the interaction of molecular quadrupole moments.

has dimension $a = b = c = 5.667$ Å. Subjected to a pressure of 3785 atmospheres or 0.38 GPa, this unit cell reduces to 5.433 Å. At low temperatures, the $\alpha$-phase can be compressed to 3500 atmospheres before it changes (to $\gamma$-$N_2$).

Tetragonal, or $\gamma$-$N_2$ exists at temperatures below 44.5 K, and between about 0.3 GPa and 3 GPa pressure. The space group of the $\gamma$ phase is P42/mnm and its unit cell has lattice constants $a = b = 3.957$ Å, $c = 5.109$ Å at 20 K and 4000 atmospheres. In the $\gamma$ form of nitrogen, the nitrogen molecules appear to have the shape of a prolate spheroid, 4.34 Å in the long dimension, and 3.39 Å in the short diameter. In addition, solid nitrogen is found to possess a $\delta$-, an $\varepsilon$-, an $\zeta$-, a $\theta$-, an $\iota$- and a $\mu$-phase depending upon the temperature and the applied pressure, which tells us that the interactions between the nitrogen molecules in the solid state are weak and easily perturbed by the applied forces of temperature and pressure. The nitrogen molecules may possess a quadrupole moment, but it is small and the molecules' oval shape mean that different orientations of the molecules in the solid are close to each other in energy, and that they readily interconvert under the influence of thermal and mechanical excitation.

In the $\alpha$-phase, figure 7.1, the nitrogen molecules are randomly tipped at an angle of 55° from the $c$-axis. The observed structure is the result of a quadrupole–quadrupole interaction between the molecules. In figure 7.1, the nitrogen molecules appear to have paired into the well-known slipped parallel orientation of quadrupolar species so as to maximize the attractive interaction generated by the intermolecular electrostatics. This is the classic *herringbone* structure seen in the structure of solid benzene (figure 2.3) and which arises by the electrostatic interaction of the quadrupole moments of the constituent molecules.

Aristotle first recorded that burning coal sometimes produced toxic fumes. Indeed, an ancient method of execution was to shut the criminal in a bathing room with smouldering coals. What was not known was the mechanism of death. The Greek physician, Galen speculated that there was a change in the composition of the air that caused harm when inhaled.

Carbon and oxygen have a total of 10 electrons in their valence shells. Following the octet rule for both carbon and oxygen, the two atoms form a triple bond, with six shared electrons in three bonding molecular orbitals. Since four of the shared electrons come from the oxygen atom and only two from carbon, one bonding orbital is occupied by two electrons from oxygen, forming a dipolar covalent bond. This causes a C ← O polarization of the molecule, with a small negative charge on carbon, and a small positive charge on oxygen. The other two bonding orbitals are each occupied by one electron from carbon and one from oxygen, forming (polar) covalent bonds with a reverse C → O polarization, since oxygen is more electronegative than carbon. In the free carbon monoxide, a net negative charge $\delta$- remains at the carbon end, and the molecule has a small dipole moment of 0.122 Debye[1].

The molecule is therefore asymmetric in terms of its electron distribution; the centre of mass of positive charge does not coincide with the centre of mass for negative charge, and there is a resultant dipole moment. By contrast, the isoelectronic dinitrogen molecule has no dipole moment.

Even though CO has a dipole moment, it is a small dipole moment (and is in the sense $(\delta-)$C–O$(\delta+)$, and the physical properties of CO mimic those of the isoelectronic $N_2$; thus the melting point of CO is 68 K (63.15 K in $N_2$), the boiling point of CO is 82 K (77 K in $N_2$). There is little to be said about these numbers other than, the complexity of the phase diagram for $N_2$ will likely be mimicked by CO, and for the same reasons of weak interactions leading to a number of orientational conformers, at similar energies, all of which are populated at low temperatures. This is what is seen, and there are many known and some predicted phases of solid CO. The structure of the lowest temperature phase of carbon monoxide is given in figure 7.2.

In its lowest temperature phase, the carbon monoxide crystal belongs to the rhombohedral lattice system (trigonal system) having the following lattice parameters: $a = b = 8.244$ Å, and $c = 11.25$ Å ($\alpha = \beta = 90°$ and $\gamma = 120°$), and the occupancy (Z) is 24. Notice in figure 7.2 how the structure maximizes the

---

[1] A Debye is defined to be one unit of charge displaced by 1Å.

(a)

(b)

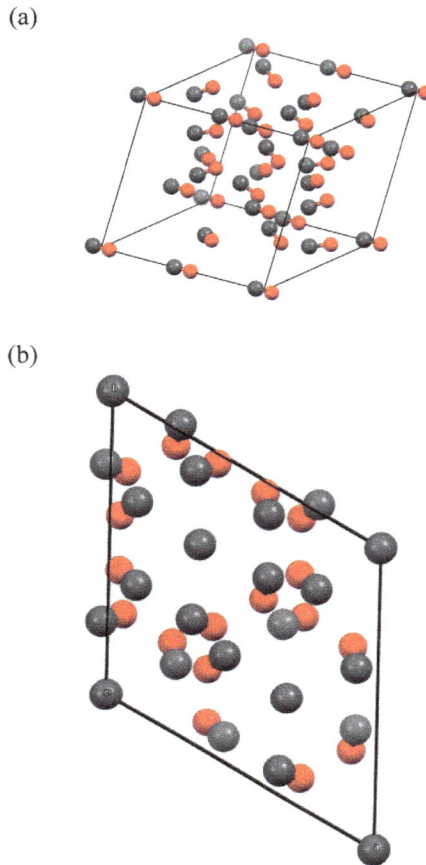

**Figure 7.2.** Two views of the structure of the lowest temperature phase of carbon monoxide (carbon atoms are grey and oxygen atoms are red).

dipole–dipole interactions between the constituent molecules; particularly, along the edge of the unit cell in figure 7.2(a) and the clusters of three molecules seen in the unit cell in figure 7.2(b). It is noteworthy that in this case there is no *herringbone* arrangement of molecules; it is a structure determined by interacting dipole moments, rather than interacting quadrupole moments.

## 7.2 Fullerenes

A fullerene is a molecule of carbon in the form of a hollow sphere, ellipsoid, or a tube. Spherical fullerenes, also referred to as Buckminsterfullerenes, resemble the balls used in soccer with a surface composed of pentagonal and hexagonal panels of carbon atoms. Cylindrical fullerenes are also called carbon nanotubes. Fullerenes are similar in structure to graphite, which is composed of stacked graphene sheets of linked hexagonal rings.

The first fullerene molecule to be discovered, and the group's namesake, Buckminsterfullerene ($C_{60}$), was manufactured in 1985 by Richard Smalley, Robert Curl, James Heath, Sean O'Brien, and Harold Kroto at Rice University. The name was an homage to Buckminster Fuller, whose geodesic domes it resembles. Fullerenes have since been found to occur in Nature, and have been detected in outer space.

As they are composed of only carbon atoms, they are all allotropes of carbon, as are diamond and graphite. However, the chemical and physical properties of fullerenes are totally different from those of diamond and graphite. This difference essentially arises because the fullerenes are a class of discrete molecules, and retain their individuality in the solid state; that is, they are not like ionic crystals or giant covalent structures. But even as discrete covalent molecules, they are only sparingly soluble in solvents such as toluene and carbon disulphide. Solutions of $C_{60}$ have a deep purple colour, as they absorb in the green, and solutions of $C_{70}$ are a reddish brown. The higher fullerenes $C_{76}$ to $C_{84}$, have a variety of colors, with $C_{76}$ possessing two optical forms; while other higher fullerenes have several structural isomers.

Solid $C_{60}$ is as soft as graphite, but when compressed to less than 70% of its volume it transforms into a super-hard form of carbon, diamond. $C_{60}$ films and solution have strong non-linear optical properties; in particular, their optical absorption increases with light intensity. Such optical properties are not unexpected given the size of the molecule and the number of electrons delocalized throughout a single $C_{60}$ molecule; that is, they are hugely polarizable and so possess a large non-linear component to their molecular polarizability (a molecular hyperpolarizability).

$C_{60}$ consists of 60 symmetry-equivalent carbon atoms with 12 isolated pentagons and 20 hexagons making the familiar soccer-ball shaped molecule (or a truncated icosahedron) with point group symmetry, $I_h$. Its electronic properties are determined by the 60 $sp^2$-hybridized carbon orbitals, which give the molecule 60 $\pi$-orbitals; that is, there are 60 delocalized electrons—an order of magnitude more than in benzene.

Given the molecules' near spherical shape, the crystal structure of $C_{60}$ reveals a face-centred cubic structure; see figure 7.3. The crystal belongs to the cubic system, with $a = b = c = 14.061(5)$ A, giving a unit cell volume of 2780 cubic angstroms; the occupancy $(Z)$ is 4. It is of interest that the diameter of the $C_{60}$ molecules is about 7 Å and the centres of the molecules are separated by about 10 Å. Attempts to refine the structure at ambient temperatures invariable reveal the presence of disorder in the lattice; indeed, in the case of $C_{60}$ disorder is seen over two distinct (but different) orientations of the molecules with occupancies of 0.714 and 0.286.

This structure therefore contains voids at its octahedral and tetrahedral sites; there are sufficient large (diameters of 6 Å and 2 Å, respectively) to accommodate impurity atoms or molecules. When alkali metals are doped into these voids, $C_{60}$ converts from a semiconductor into a conductor, or even to a superconductor (see the following section).

Another manifestation of the perturbation of $C_{60}$ molecules by organic solvents of polarizable molecules is that $C_{60}$ can crystallize with some solvent molecules caught up in the crystal lattice (these structures are termed solvates, and arise because of the strong intermolecular interactions that occur between solvent molecules such as

(a)

(b)

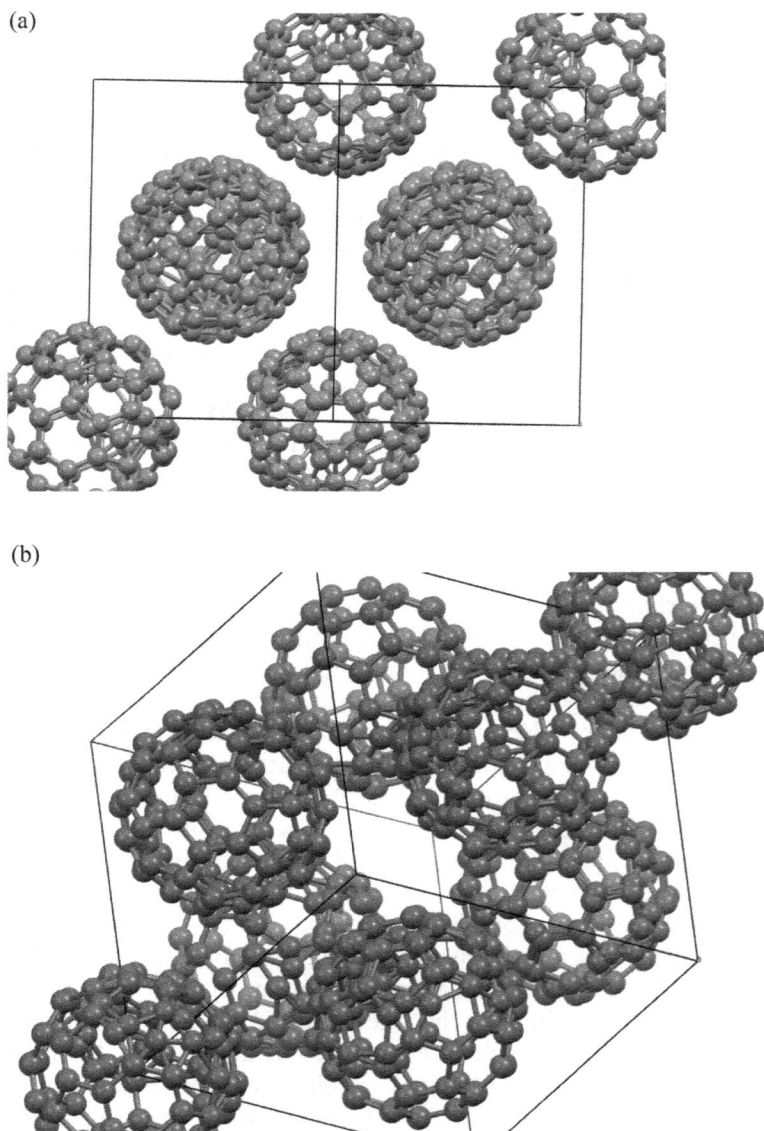

**Figure 7.3.** Two views of the structure of $C_{60}$. The main difference between these two images is the number of molecules included in the viewing range of the visualization software. The radius of the $C_{60}$ molecule is about 3.55 Å, and the distance between the centres of two molecules is about 10 Å. See http://www.chemtube3d.com/gallery/structurepages/c60.html for an interactive, animated structure.

benzene and the $C_{60}$ molecule which has many faces made up of $C_6$ units as is benzene, and so there is a particularly strong intermolecular interaction). For example, crystallization of $C_{60}$ in benzene solution yields triclinic crystals with the formula $C_{60}(4C_6H_6)$. Like other solvates, this one readily releases benzene on heating to give the usual face-centred cubic $C_{60}$.

In solid Buckminsterfullerene, the $C_{60}$ molecules are held together via weak van der Waals forces in the face-centered cubic arrangement. In figure 2.3 we saw the structure of benzene and how it could be rationalized by the interacting molecular quadrupole moments, in $C_{60}$ the molecules are ten times more polarizable, but they are spherical and so the intermolecular forces are similar but generate very different structures. At low temperatures the individual molecules are locked against rotation in the solid. Upon heating, however, they start rotating at about $-20$ °C. This results in a first-order phase transition at 260 K to a face-centred cubic structure and a small increase in the lattice constant from 14.11 Å to 14.154 Å. A similar behaviour is seen in solid benzene, where at about 150 K the benzene molecules begin to undergo librational, or rotation-like reorientation in the solid, but without disturbing the crystal structure, but benzene melts at 278.8 K

The structural phase transition at 260 K concerns a change of the reorientational (rotation-like) motion of the $C_{60}$ molecules in the crystal, below 260 K they become slower and better described as a librational motion (localized or bound motion—not a free rotation). The phase transition is accompanied by a contraction of 0.34% in the cubic lattice constant. Reflections not compatible with face-centred cubic symmetry are now present in the powder diffraction patterns, reflecting a long-range orientational ordering of the fulleride units.

Refinement of neutron powder diffraction data from low temperature measurements has identified two types of C–C bonds related to short *double bonds* (1.40 Å) and long *single bonds* at 1.45 Å. The double bonds, also known as 6:6 bonds, fuse two hexagons together, whereas the longer hexagon:pentagon fusions are known as 6:5 bonds. The experimentally determined structure comprises a majority fraction of molecules rotated by $\varphi \approx 98°$ from the so-called standard orientation, so that optimization of intermolecular interactions occurs with electron-rich double bonds lying over electron-deficient pentagonal faces of neighbouring molecules (the slipped parallel arrangement seen in benzene in figure 2.2). A co-existing minority orientation is only slightly less energetically favorable (by ~11 meV) and is characterized by the alignment of 6:6 bonds parallel to hexagonal faces of neighbours with a rotation angle of $\varphi \approx 38°$ from the standard orientation.

On cooling from 260 K to ~90 K, the fraction of molecules in the majority 98° orientation increases from roughly 60% to about 83.5%, whereas below 90 K the fraction of molecules in the 98° orientation remains constant, because the molecules now have insufficient thermal energy to overcome the potential barrier separating the two orientations. A cusp in the rate of change of lattice constant with temperature is observable at the same temperature. At this temperature, the dynamics of the $C_{60}$ molecules are limited by the lack of thermal excitation, and the molecules are only able to perform small amplitude librational motion.

## 7.3 Alkali-metal fullerides and superconductivity

In 1991, it was discovered that intercalation of alkali-metal atoms into solid $C_{60}$ leads to metallic behaviour. It was seen that potassium-doped $C_{60}$ becomes superconducting at 18 K. This was the highest transition temperature for a molecular superconductor (cuprate oxide materials become superconducting at much higher temperatures, but they have more complex mixed bonding). Since then, superconductivity has been reported in fullerene doped with various other alkali metals. It has been shown that the superconducting transition temperature in alkaline metal-doped fullerene increases with the unit-cell volume. The highest superconducting transition temperature of 33 K at ambient pressure, is reported for $Cs_2RbC_{60}$.

The increase of transition temperature with the unit-cell volume had been suggested to be evidence for the BCS, or Bardeen–Cooper–Schrieffer (named after John Bardeen, Leon Cooper, and John Robert Schrieffer) mechanism of alkali metal-$C_{60}$ solid state superconductivity, because inter-$C_{60}$ separation can be related to an increase in the density of states on the Fermi level. Therefore, there have been many efforts to increase the inter-fullerene separation, in particular, intercalating neutral molecules into the $A_3C_{60}$ lattice to increase the inter-fullerene spacing while the valence of $C_{60}$ is kept unchanged.

However, it is not only via the BCS mechanism that one may explain the origin of high-temperature superconductivity in alkali metal fullerides. It is possible to present a different mechanism that generates an attractive force between similarly-charged particles, which could lead to the possible condensation of Cooper pairs into a boson-like state.

There is considerable interest in the properties of voids in crystal structures; voids in zeolites have catalytic properties, clathrates are ephemeral crystals held together by hydrogen bonds that exist only under extreme conditions, and alkali metal fullerides have unique optical and electrical properties, unrelated to the separate alkali metals and $C_{60}$ molecules. We will see below how the presence of molecules in large cavities or voids in crystals can profoundly modify the Coulombic ion–ion repulsive interaction to be expected in ionic materials. Indeed, we will outline a theory to explain the superconductivity that is seen in fullerides.

Figure 7.4 gives the structure of $Cs_2RbC_{60}$, but for simplicity the two different alkali metals are not distinguished. It is seen that the metal atoms (in purple) are located in the voids between close-packed $C_{60}$ molecules (shown in grey). The structure is cubic with a unit cell characterized by the length 14.555(7) Å; the unit cell has a volume of 3083.45 cubic angstroms and the occupancy ($Z$) is 4. There is disorder in this structure associated with the relative orientations of the $C_{60}$ molecules, as is seen in pure $C_{60}$; there are two possible orientations present in the solid.

In what follows, we will consider the metal atoms and molecules to be ionized; that is, existing as $K^+/Cs^+$ and $Bf^-$, which generates a typical ionic material (cf. sodium fluoride), built of ions where the valence is $C_{60}^{3-}$. The electrons have been transferred from the metal atoms to the walls, or the surfaces of the $C_{60}$ molecules; that is, to the walls of the cavities within the cubic structure. This is a textbook example of Gauss' theorem; where the potential inside a closed volume arising from charge on its surface is zero. There is therefore no net ion-pair dipole of the cation with the localized electron/host molecule, because an image charge can always be found on the opposite side of the cavity for nullification.

(a)

(b)

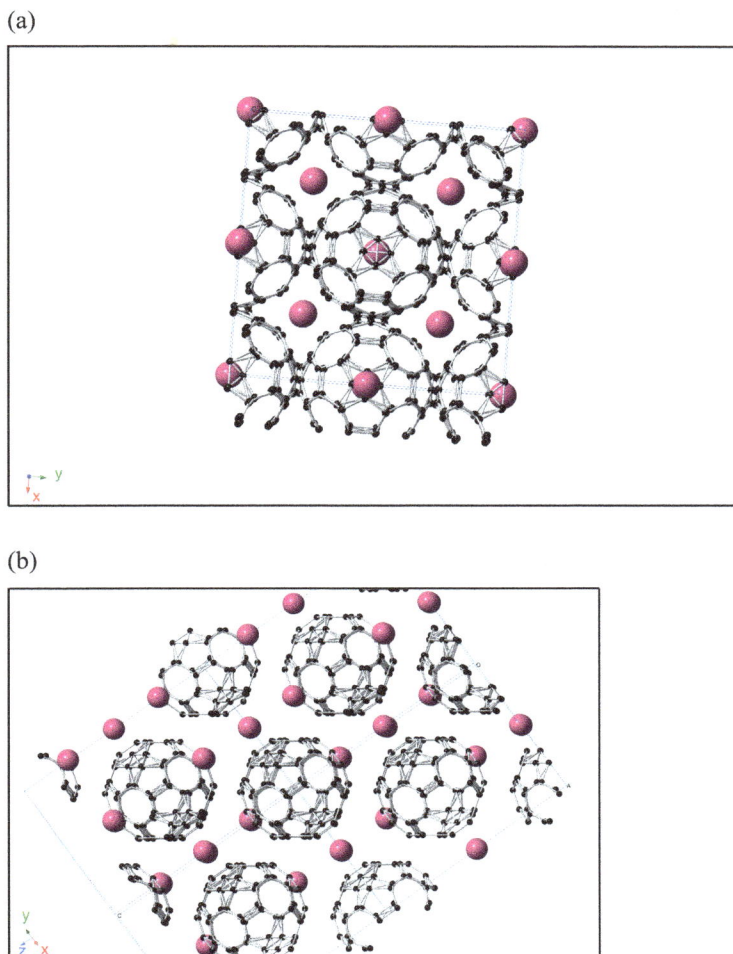

**Figure 7.4.** Two views of the structure of the alkali-metal fulleride $Cs_2RbC_{60}$, where the two different alkali metals are not distinguished, and are shown as purple spheres. See http://www.chemtube3d.com/solidstate/SS-C60K.htm for an interactive animated structure.

Consequently, in these systems we have bare cations inside large voids where they are not subject to any restoring potential arising from the transferred electrons. Such a system will be particularly susceptible to any additional electrical fields within the solid. If these fields are frequency dependent, then the cations will be driven at the frequency of the oscillating fields. As the cations are free to move within the void, they will have a large dynamic electrical susceptibility, or vibrational polarizability. And this dynamic electric polarizability can profoundly alter the electrical properties of the solid.

Consider the following model. The cation–cation distance is $R_{++}$ and the cation-$C_{60}$ distance (to the surface of the $C_{60}$ molecule) is $R_{+-}$; we will assume that the cation is in an harmonic potential with regard the surface of the cavity and is perturbed by the phonons (of frequency $\omega$) of the $C_{60}$ lattice. The dynamic polarizability, $\alpha(\omega)$, of the cation can then be written in terms of the excitation frequencies of the cation, $\omega_0$:

$$\alpha(\omega) = e^2/m_+(\omega_0{}^2 - \omega^2), \tag{7.1}$$

where $e$ is the electronic charge and $m_+$ is the mass of the cation. Although a permanent ion-pair electric dipole will not exist in such solids due to symmetry, there will be a fluctuating distortion induced dipole moment arising from thermal distortions of the host lattice; that is, there will be an instantaneous dipole ($\mu$) proportional to the mean amplitude of motion of the lattice, $< u >$, and the electric charge, $\mu = e. < u >$.

The dynamic polarizabilities can be large. Powder diffraction measurements [1] provide values of $< u >$, it is the Debye Waller factor, and the lowest frequency phonon of $K_3C_{60}$ is 104.8 cm$^{-1}$ [2], and calculation [3] gives $\alpha(\omega)$ for a $K^+$ ($\alpha_{K+}$) in an octahedral site of the $C_{60}$ lattice to be about $300 \times 10^{-40}$ C$^2$m$^2$J$^{-1}$ and $-2000 \times 10^{-40}$ C$^2$m$^2$J$^{-1}$ for a K+ in a tetrahedral site (at tetrahedral sites the amplitude of motion is much larger than at octahedral sites). The electronic polarizability of $C_{60}$ is also large at $108 \times 10^{-40}$ C$^2$m$^2$J$^{-1}$ [3] (about ten times the magnitude of the polarizability in benzene).

The presence of these large dynamic polarizabilities of the cations will significantly change the nature of the interaction between the cations and the host. And the large polarizability of the host molecules will further distort the interactions within the fulleride.

The distortion of the large polarizability of $C_{60}$ is a possible means of overcoming the classical Coulombic repulsion of the cations in the interconnected voids of the host (see the structure in figure 7.4). The cation–cation repulsion energy is

$$U(rep) = e^2/(4\pi\varepsilon_0)R_{++} \tag{7.2}$$

Consider one of the cations; it will polarize the $C_{60}$ molecules around it and through the polarizability of the $C_{60}$ molecules induce a dipole, $\mu_{C60}$. Thus

$$\mu_{C60} = \alpha_{C60}. E_{(cation)} = \alpha_{C60}. \{e/(4\pi\varepsilon_0)R_{+-}{}^2\} \tag{7.3}$$

The electric field of this dipole will then be able to interact with a second cation. This field is

$$E_{(C60)} = \mu_{C60}/(4\pi\varepsilon_0)R_{+-}{}^3, \tag{7.4}$$

the electrostatic interaction of this field together with that of the first cation on the polarizability of the second cation is termed the polarization energy, and is represented as

$$U_{(pol)} = \tfrac{1}{2}\sum \alpha E_j^{(0)} E_j, \tag{7.5}$$

where $E_j$ is the field due to the polarized $C_{60}$ molecule and $E_j^{(0)}$ is the field due to the first cation. The polarizability of the cations is $\alpha(\omega)$, and the sum runs over all nearest neighbour polarized $C_{60}$ molecules; $\alpha(\omega)$ for an octahedral located potassium cation is about $300 \times 10^{-40}$ $C^2m^2J^{-1}$. Thus, for our trio of two potassium cations and an intermediate $C_{60}$ molecule,

$$U_{(pol)} = (\tfrac{1}{2})\alpha_{K+}\{e/(4\pi\varepsilon_0)R_{++}^2\}\{\mu_{C60}/(4\pi\varepsilon_0)R_{+-}^3\}. \qquad (7.6)$$

X-ray diffraction tells us that for $K_3C_{60}$, $R_{++} = 6.17$ Å and $R_{+-} = 3.69$ Å. These values give a repulsive Coulombic interaction between two cations as about 2.34 eV, and for the attractive, polarization energy 6.87 eV. For interactions of two neighbouring octahedral sites, the sum in equation (7.5) would run over two possible interactions, and for the interaction of tetrahedral with octahedral sites, the sum would run over three possible interactions.

The presence of polarizable matter between polarizable, charged particles (of like charge) can give rise to an attractive interaction much stronger that the repulsive Coulombic interaction to be expected. This mechanism is reminiscent of that responsible for Cooper pairs of electrons in the BCS theory for the origin of superconductivity. Here, however, the phonons of the host produce a dynamical polarizability of the relatively light intercalated cations, which is able to interact with the polarizable matter between the cations via an electronic distortion.

It is worth pointing out that as the temperature of the fulleride changes, the dimensions within the solid will change, and so the distances between the ions will vary with temperature. And it is likely that the relative magnitudes of the attractive and repulsive interactions between the cations will also change with temperature. However, the repulsive term between the cations being linear in $R_{++}$ will be less affected by thermally driven changes in the interionic spacings than will be polarization energy, which is dependent on the product $R_{++}^2.R_{+-}^3$.

## Further reading

[1] Stephens P W, Mihaly L, Lee P L, Whetten R L, Huang S-M, Kaner R, Deidevich F and Holczer K 1991 *Nature* **351** 632
[2] Prassides K, Dennis T J S, Hoare J P, Tomkinson J, Kroto H W, Taylor R and Walton D R M 1991 *Chem. Phys. Lett.* **186** 455
    White J W, Lindsell G, Pang L, Palmisano A, Sivia D S and Tomkinson J 1992 *Chem. Phys. Lett.* **191** 92
[3] Williams J H 1992 *Physica* B **182** 261–6

**IOP** Concise Physics

# Crystal Engineering
How molecules build solids
**Jeffrey H Williams**

# Chapter 8

# Methane and other non-aromatic hydrocarbons: ethane, ethylene and acetylene

Not all atoms are alike, and it is through the differences in atoms that the electrical properties of the molecules they construct are given form. Certain atoms, such as fluorine and oxygen, have a propensity (electronegativity), when covalently bonded with other atoms for pulling the shared electrons into their structure. Fluorine has the highest electronegativity and hydrogen has a low electronegativity. Thus, a fluorine atom when chemically bonded to another atom will pull a shared electron from the other atom towards itself. The consequence of such charge asymmetries in molecules is that there is a slight electrostatic attraction between two or more such molecules. This type of weak electrostatic interaction is manifest in more energy being required to separate individual molecules from bulk quantities of those molecules.

Depending upon the shape of the molecules, the charge asymmetry of the covalent bonds in the molecules may lead to permanent electric moments. There are electric dipoles in CO, and quadrupole moments in $N_2$. Ongoing from a linear shape to a tetrahedron and on to an octahedron and then to a sphere, the electrical moments become increasingly smaller and smaller, vanishing in a sphere; in chapter 2, we saw how the symmetry of a molecule can tell us a great deal about the electrical properties of that molecule.

After the quadrupole moment, the next smallest electrical moment of a molecule is the octupole moment, which would be one unit of electric charge spread throughout a volume, about $3.33 \times 10^{-50}$ C m$^3$ in the SI. Molecules possessing an octupole moment, arising from the asymmetry in the distribution of electronic charge within their structure, are tetrahedral in shape, for example methane[1].

---

[1] In terms of a pair of interacting molecules, what the absence of a dipole or a quadrupole moment means is that the energy of interaction is very small. Water is a strongly associated fluid, containing a strong dipole–dipole interaction and hydrogen bonds, and as a consequence boils at 100 °C. Benzene has no dipole moment and no hydrogen bonds but has a quadrupole moment, and boils at 80 °C. Methane has no quadrupole moment and only an octupole moment, and as a consequence boils at −161.5 °C, and freezes at −182.5 °C.

doi:10.1088/978-1-6817-4625-8ch8     8-1

When we arrive at these symmetric molecules, we are close to the limits of modelling intermolecular forces in terms of molecular multipole moments, and must invoke the dispersion forces as means of rationalizing the attractive forces required for condensation. But when we are considering a neon atom, there is so much symmetry (it is a sphere), and its orientation within a solid lattice is meaningless. Similarly, when we come to solid methane we are close to the limits of being able to talk about the orientation of one molecule relative to a neighbouring molecule in a crystal lattice, because there are so many possible orientations and they all have a similar energy, and so are all populated at a particular temperature (even a low temperature). And this means that there is orientational disorder within the crystal at these low temperatures. Crystals form, and these crystals may be subjected to x-ray diffraction, but more than one structure, containing different relative orientations of neighbouring methane molecules may be fitted to the observed data. These structures are closely related but are not identical, and this leads to uncertainty in the fitting process, thereby limiting the ultimate precision of the measurements, and of the determined structure.

Methane is the simplest organic molecule, but like many other supposedly simple molecules it has a rich and complex phase diagram. Of at least seven known phases of solid methane, only the structures of the two cubic phases have been completely solved. In both cases, the carbon atoms occupy a face-centred cubic lattice, demonstrating a tendency of the methane molecules to approximate to spheres and to form close-packed crystal structures. In phase I, which at ambient pressure is stable below a melting temperature of about 90 K and above $T = 20.4$ K in $CH_4$ and $T = 27.0$ K in $CD_4$, all of the tetrahedral molecules are orientationally disordered. In phase II, below these temperatures, the orientation-dependent octupole–octupole interaction leads to partial orientational order. The crystal structure is described by the space group *Fm3c*, with six orientationally ordered sublattices and two disordered sublattices; it is sometimes referred to as antiferrorotational.

The structure of the lowest temperature phase of perdeuteromethane ($CD_4$) is given in figure 8.1. This crystal belongs to the tetragonal system, and has the following unit cell dimensions, $a = b = 11.537$ Å, and $c = 11.723$ Å, and the occupancy ($Z$) is 32. This structure was solved by neutron diffraction. Neutrons are scattered by nuclei, whereas x-rays are scattered by electrons, but the neutron scattering cross section of an atom which determines the atom's ability to scatter neutrons consists of two terms, an incoherent part and a coherent part. And as the name suggests it is the coherent part that is important in structure determination via Bragg's law (inelastic and quasielastic scattering on neutrons depends upon the incoherent neutron scattering cross section)[2].

As can be seen in figure 8.1, there are 32 molecules per unit cell. With so many weakly-interacting molecules in the unit cell, the scope for orientational disporder is

---

[2] The incoherent and coherent scattering cross sections for the hydrogen atom are 80 and 1.8 barns, respectively, and for the deuterium atom they are 2 and 5.6 barns, respectively; 1 barn = $10^{-24}$ cm$^2$. Consequently, for structural studies using neutron scattering, one is advised to use deuterated molecules. But for studies of molecular dynamics using neutron scattering, one uses hydrogenated materials.

(a)

(b)

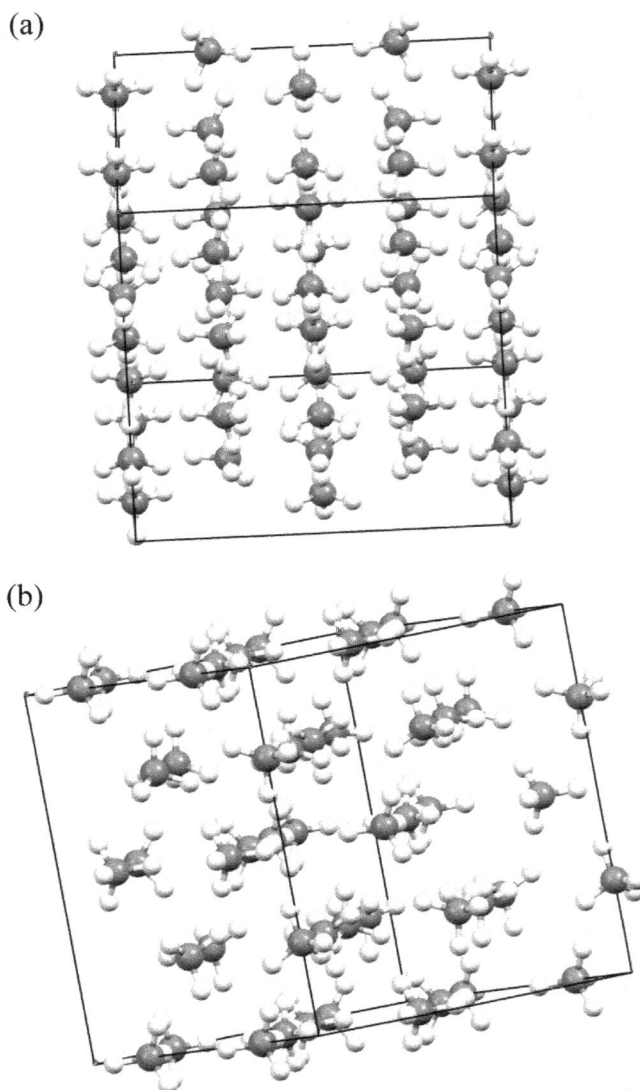

**Figure 8.1.** Two views of the structure of the lowest temperature phase of solid per-deuteromethane ($CD_4$); the carbon atoms are dark grey and the hydrogen/deuterium atoms are white.

significant. The weakness of the intermolecular interactions; essentially dispersion forces (varying as $r^{-6}$) and a weak octupole–octupole interaction (varying as $r^{-7}$) is seen in the low boiling temperature of this material.

After methane, ethane ($C_2H_6$) is the next simplest hydrocarbon. And like methane, ethane is one of those materials that people only think of as a gas, but like methane (and indeed every other gas) given cold enough conditions ethane will freeze to a solid. Ethane freezes at 90 K, where it becomes a plastic crystal, meaning that the molecules within the structure are freely rotating; that is, there is significant orientational disorder. This means that the structure revealed by x-ray diffraction is deceptively simple; a closely packed set of spheres in a body-centred arrangement. However, this plastic phase has only a very narrow range of stability. When this phase is cooled from 90 K to 89.5 K, the molecules cease rotating, and the structure becomes monoclinic; that is, much less symmetric than when it first freezes.

The plastic modification of crystalline ethane, at 90 K, is cubic with $a = b = c = 5.304$ Å and $Z = 2$. At a slightly lower temperature, ethane has transformed into a monoclinic form with a unit cell of: $a = 4.226$ Å, $b = 5.623$ Å and $c = 5.845$ Å, and $\beta = 90.41°$; this monoclinic form has an occupancy of 2. In the monoclinic phase, the C–C bonds are fixed (with a mean amplitude of motion of $<u_C^2> = 0.031$ Å$^2$), and preferred orientations of the hydrogen atoms may be inferred.

The arrangements of the ethane molecules in the crystalline monoclinic form given in figure 8.2 are reminiscent of the orientation of other quadrupole molecules. It is a classic herringbone arrangement of molecules (see figure 2.3 for the same arrangement of molecules in solid benzene and figure 7.1 for the same structure in solid nitrogen). The two methyl groups of the ethane molecule can rotate around the C–C axis and so a number of conformers are possible, ranging from staggered to eclipsed; their presence will be temperature dependent.

The two possible conformations of the methyl groups in ethane are, 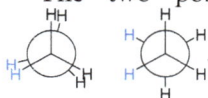. The first form is the eclipsed form and the second is the staggered form. These two forms arise by rotation around the C–C bond of the molecule. When the two methyl groups in ethane are neither staggered, nor eclipsed, the molecule has $D_3$ symmetry. When the methyl groups are eclipsed, the symmetry of the molecule is $D_{3d}$, and when the methyl groups are staggered, the symmetry of the molecule becomes $D_{3h}$. These symmetry changes will profoundly alter the vibrational spectroscopy of solid ethane; the infrared spectrum will become a function of temperature. They will also influence the attractive electromagnetic forces between the molecules in the solid at these low temperatures.

One conformer will have a centre-of-inversion and so may not possess an electric dipole moment, but will have a quadrupole moment; and the structure of the solid will be driven primarily by the quadrupole–quadrupole interaction varying as $r^{-5}$ (where $r$ is the intermolecular separation). The other conformer will not have a centre-of-inversion and so will have a small dipole moment in addition to its quadrupole moment. And this addition term, varying as $r^{-3}$, will influence the intermolecular forces seen in this system. The quadrupole moment of ethane is $-2.7 \times 10^{-40}$ C m$^2$ (the phase tells us that the charge is mainly located at the extremities of the molecules—on the hydrogens). The interaction of such small quadrupole moments demonstrates why ethane is a gas that boils at 185 K (methane boils at 112 K).

The related hydrocarbon ethylene, $C_2H_4$ is the simplest hydrocarbon containing a C=C double bond, or two bonded $sp^2$-hybridized carbon atoms. The presence of a

(a)

(b)

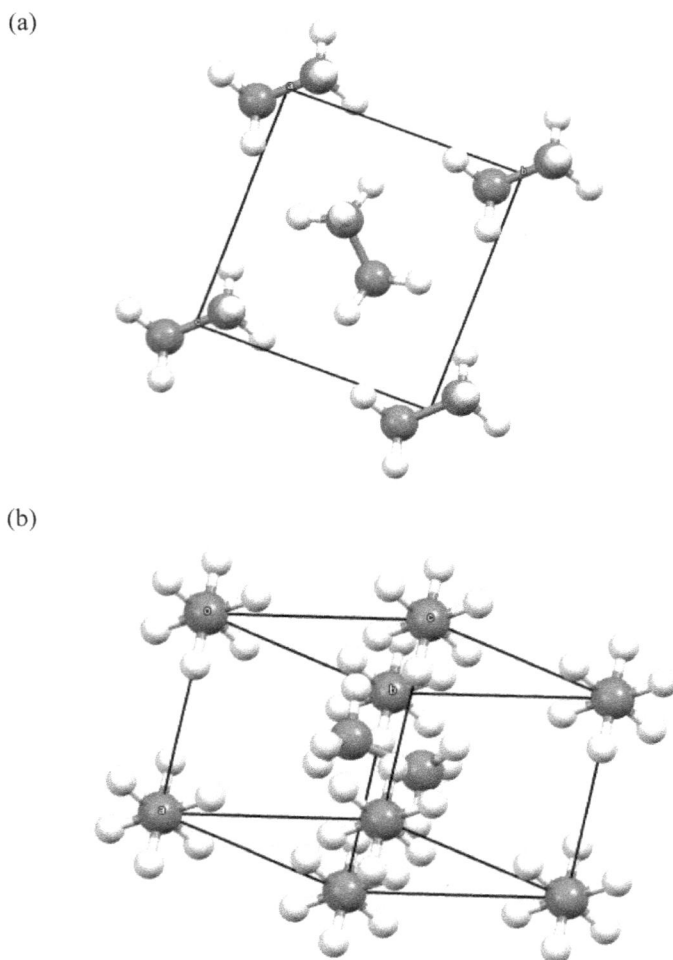

**Figure 8.2.** Two views of the structure of ethane in its crystalline phase (carbon atoms are dark grey and hydrogen atoms are white). Both the plastic and monoclinic phases, were discovered in 1978 van Nes and Vos [1].

$\pi$-bond between the two carbon atoms prevents free rotation about the axis running parallel to the C=C bond; and so ethylene has $D_{2h}$ symmetry (permanently).

The structure of solid ethylene is given in figure 8.3, and it is again seen that the crystal structure displays a classic herringbone form. The arrangements of the ethylene molecules in the monoclinic form given in figure 8.3 is reminiscent of the orientation of other quadrupole molecules (see figure 2.3 for the same arrangement of molecules in solid benzene and figure 7.1 for the same structure in solid nitrogen). This material belongs to the monoclinic system and has the following unit cell dimensions: $a = 4.626$ Å, $b = 6.620$ Å, $c = 4.067$ Å, with $\beta = 94.39°$ and an occupancy ($Z$) of 2 molecules per unit cell.

With ethylene, because of the symmetry of the molecule (the main rotational axis of the molecule is $C_2$, rather than $C_n$ where $n \geqslant 3$) we are no longer able to speak of a single value of the quadrupole moment. That is, the charge distribution of this non-polar molecule cannot be defined by a single number. In a molecule such as $CO_2$ and acetylene, $\Theta_{zz} = -(\Theta_{xx} + \Theta_{yy})$ and $\Theta_{xx} = \Theta_{yy}$; in ethylene, however, $\Theta_{zz} = 6.6 \times 10^{-40}$ $C\,m^2$, $\Theta_{xx} = 6.9 \times 10^{-40}\ C\,m^2$ and $\Theta_{yy} = -13.5 \times 10^{-40}\ C\,m^2$.

In acetylene, $C_2H_2$, we arrive at a molecule that has a distinctive shape (linear) and charge distribution that leads to a structure less susceptible to orientational disorder and plasticity. This is the simplest molecule containing a C–C triple bond; that is, composed of two carbon atoms bonded by *sp* hybrid orbitals. The molecule is linear centro-symmetric; the symmetry is $D_{\infty h}$ and so the molecule may not possess a dipole moment, and the quadrupole moment of acetylene is $24 \times 10^{-40}\ C\,m^2$. This is a large value (close to that of hexafluorobenzene at $31 \times 10^{-40}\ C\,m^2$) and reflects the presence of the triple bond (a $\sigma$-bond and 2 $\pi$-bonds) between the carbon atoms; the centre of the charge is off-axis between the two carbon atoms. And the phase of the value of the quadrupole moment of acetylene tells us that the two hydrogen atoms carry a $\delta+$ charge and that the bulk of the electron distribution is off-axis and in the middle of this linear molecule; $\delta(+)$H–C–$(2\delta-)$–C–H$(\delta+)$.

The structure of solid acetylene is seen in figure 8.4. These structures belong to the cubic system, and so $a = b = c = 6.094$ Å, and there are four molecules per unit cell.

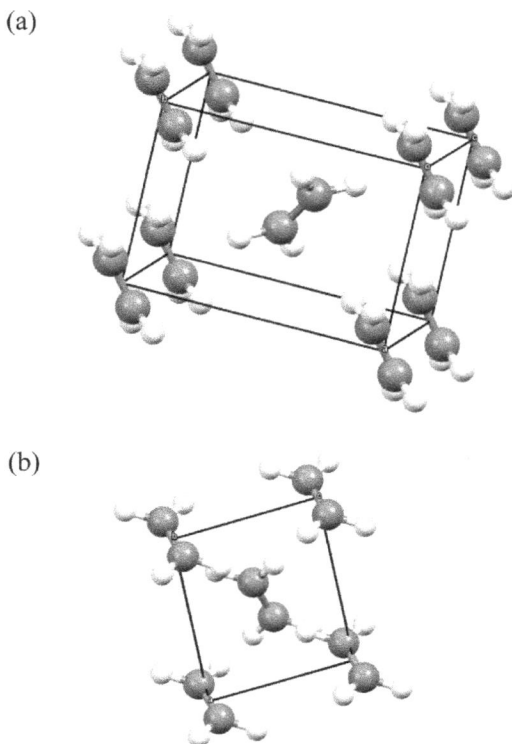

(a)

(b)

**Figure 8.3.** Two views of the structure of ethylene; the carbon atoms are dark grey and the hydrogen atoms are white.

(a)

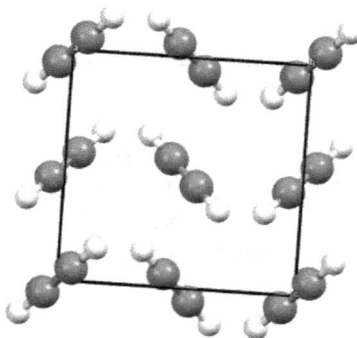

(b)

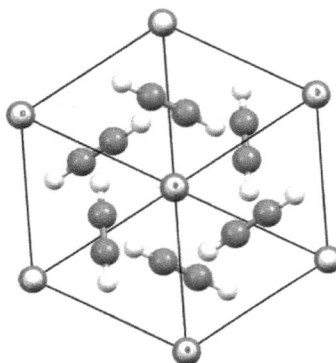

**Figure 8.4.** Two views of the structure of solid acetylene (the carbon atoms are dark grey to black and the hydrogen atoms are white).

Given the size of the molecular quadrupole moment it is no surprise that the structure of the solid is characterized by the usual slipped parallel and perpendicular or T-shaped arrangements of quadrupole moments of like phase; that is, we have a structure reminiscent of solid benzene, solid nitrogen and solid ethylene. The classic herringbone structure of solid acetylene is clearly seen in figure 8.4(a). And figure 8.4(b) shows the hydrogen atoms of acetylene molecules (which can be given a partial charge of $\delta+$) attempting to point directly at the electron density off-axis in neighbouring acetylene molecules; that is, point at the cloud with partial charge $\delta-$. Of course, the ideal slipped parallel and T-shape arrangement of quadrupole moments seen in planar benzene is distorted by the linear shape of these molecules.

## 8.1 Disorder in organic crystals

Given that the intermolecular bonding in organic crystals is relatively weak, certainly weak when compared to ionic crystals, there is often some difficulty in generating a detailed structure from x-ray diffraction measurements. Disorder is frequently found to be present in organic crystals. Indeed, about 20% of all structures in the Cambridge Structural Database (https://www.ccdc.cam.ac.uk/) report the presence of some kind of disorder in the derived crystal structure. This disorder can take many forms; for example, the molecules may be in motion at the temperature of the measurement. Such dynamic disorder can arise from thermally-driven motion of the molecules within the solid; for example, the energy difference between one orientation of a molecule in the unit cell may differ from another orientation by an amount of energy that is, of order, the available thermal energy, in which case there may be a number of different orientations in the crystal (we will see later that the adduct formed between 1,3,5-trimethylbenzene and hexafluorobenzene undergoes solid-state phase transitions related to the ordering of the methyl groups). Alternatively, different constituent molecules may occupy a continuum of different energy positions (this is static, but continuous disorder), or each constituent may occupy one of a limited number of possible states, or *disorder components* (static, discrete disorder). Examples of this last kind include disorder in species substitution and orientational disorder where the molecule is restricted to a few, well-defined, orientations.

For crystallographers, a disordered system typically results in a diffraction pattern that refines to, or generates a crystal structure in which each site, or sites is occupied by one of two (or more) possible 'components', with some approximately known probability; for example, the site is occupied by molecule A or by molecule B with a 50:50 probability. In the presence of disorder, all that can be inferred is that within some large volume of the crystal, half of the total sites are occupied by A, and half by B. Any probability is possible in principle and it could well be temperature dependent (certainly true when the local orientational minima on the intermolecular potential are close to $k_B T$). However, the true pattern in the distribution of A and B (on a super-unit cell scale) is difficult to determine, and a full characterization of the crystal is not possible. Methods for inferring short-range ordering in disordered systems based on diffuse scattering are available, but require high-quality data.

Solids that are composed of near-spherically symmetric molecules; for example, $C_{60}$ or methane are able to occupy many possible lattice positions even at low temperatures when the energy available to populate higher librational and vibrational position within a crystal is low. In such rotator phases, there are many possible sites on the potential energy surface seen by molecules in the solid; that surface is flat and isotropic, which means that even at low temperatures the molecules are undergoing large amplitude motions. These motions may involve the whole molecule rattling about in its lattice position, or it may be part of a molecule moving relative to the rest of the molecule that is in a fixed position. The Debye–Waller factor is a means of accounting for the influence of the temperature-driven motion of the scattering atoms and molecules on the intensities of the Bragg peaks.

The Debye–Waller factor (DWF), named for Peter Debye and Ivar Waller, is used in condensed matter physics to describe the attenuation of x-ray scattering and coherent neutron scattering caused by thermal motion of the scattering atoms. The DWF depends on the scattering vector $\mathbf{q}$. In diffraction studies, only the elastic scattering is useful; in crystals, it gives rise to distinct Bragg peaks. Inelastic scattering events are undesirable as they cause a diffuse background scattering, as seen in the highest temperature diffraction pattern in figure 3.2.

The basic expression for the DWF is given by

$$\text{DWF} = \langle \exp(i\mathbf{q}. \mathbf{u}) \rangle^2,$$

where $\mathbf{u}$ is the displacement of a scattering centre, and the angle brackets denotes either thermal or time average. Assuming that the displacements of the atoms from their mean positions in the crystal are harmonic oscillations, one may write

$$\text{DWF} = \exp(-q^2 <u^2>/3)$$

where $q$ and $u$ are the magnitudes (scalar values) of the vectors $\mathbf{q}$ and $\mathbf{u}$, respectively, and $<u^2>$ is the mean squared displacement. Note that if the incident wave has wavelength $\lambda$, and it is elastically scattered into an angle of $2\theta$, then $q = 4\pi \sin \theta/\lambda$.

In structural investigations of very large molecules, the term $B$-factor is often used; defined as $B = 8\pi^2 <u^2>$. The $B$-factors can be taken as indicating the relative vibrational motion of different parts of the structure under investigation. Atoms with low $B$-factors belong to a part of the structure that is well ordered. Atoms with large $B$-factors generally belong to part of the structure that is very flexible and often display orientational disorder.

The DWF does not shift the position at which monochromatic x-rays are scattered by a thermally-excited structure, but it does decrease the precision with which the diffraction angles may be located, and hence the precision with which the structure may be determined from the measured data.

## 8.2 Thermal diffuse scattering

Thermal diffuse scattering of x-rays refers to scattering generated by the motion of the object doing the scattering, but generally refers to motion of larger amplitude than is being described by the DWF. It thus provides a means of looking at the dynamics (vibrations and rotations) of the molecules that constitute the lattice, and how these localized molecular vibrations and rotations couple to generate lattice thermal vibrations, i.e. phonons. Unlike Bragg diffraction which is characterized by sharp peaks in $k$-space, thermal diffuse scattering is diffusely distributed due to the continuous distribution of phonon modes. Indeed, for crystallographers interested in Bragg diffraction, thermal diffuse scattering is often considered a nuisance, or a background that must be removed from the data[3].

Whereas Bragg diffraction occurs when scattering amplitudes add coherently, if there is a defect in a crystal lattice (e.g. atom missing, or in a slightly *wrong* position due to thermal excitation), then the amplitude of the Bragg peak decreases. This *lost* scattering intensity is redistributed into diffuse scattering. The diffuse scattering thus arises from the local (short-range) configuration of the material (not the long-range

---

[3] The neutron diffraction data measured at 285 K and displayed in the top panel of figure 3.2 displays a particularly good example of thermal diffuse scattering (this can be seen again in the same material as lower temperatures in figure 11.4(b). The sharp Bragg lines appear to grow out of a broad background feature. This broadening arises from the large amplitude motion of the scatters (see below).

structural order). In a measured diffraction pattern, the diffuse scattering is observed as a broad feature upon which is superimposed the sharp coherent Bragg lines.

In the limit of total disorder in the crystal structure (a glass), one lacks a lattice, and so the scattering does not generate any Bragg peaks. However, a disordered structure will still give rise to diffuse scattering. The Fourier transform of a disordered structure will not give any well-defined peaks, but will give a distribution of scattering intensity over a wide-range of angles. Thus, samples with an inherently disordered structure (dynamic disorder in molecular crystals, randomly packed nanoparticles, free rotation of molecules within crystalline structures, etc) will only generate diffuse scattering.

The effect of the thermal motion in a crystal on a beam of x-rays traversing that crystal may be compared with the effect of the agitated surface of the sea on the image of the setting or rising sun. There is no sharp reflection, but a diffuse ribbon of light stretching to the observer from the horizon. This diffuse scattered light is generated by innumerable waves of various lengths and directions.

In figure 8.5(a) we see the structure of Phase II of the binary adduct benzene: hexafluorobenzene; this phase exists between 247.5 and 272 K (some of the raw data used for determining this structure is seen in figure 3.2). The *c*-axis is running vertically parallel to the axis of the column of alternating equi-spaced benzene and hexafluorobenzene molecules. In this particular structure determination, the software has found significant motion in the fluorine atoms of the hexafluorobenzene molecules, and this is represented in the structure by large thermal ellipsoids, which give a measure of the uncertainty in position caused by the thermal motion of the molecule. This motion may be seen in figure 8.5(b), where we see the thermal diffuse scattering as a streak that distorts the position of a particular Bragg line. What this association of the thermal diffuse scattering with a particular Bragg reflection tells us is that the motion giving rise to this diffuse scattering comes from the almost free-rotation of the benzene molecules in this phase of the solid; that is, the benzene molecule, which lies in the *ab*-plane in the crystal is free to execute large amplitude jumps (larger than those observed in the heavier hexafluorobenzene molecules). It is the fact that the observed thermal diffuse scattering is only observed with a particular set of reflections that informs us the molecular motion in this phase originates in a plane perpendicular to the *c*-axis; that is, the near free-rotation of the benzene rings and the large amplitude motion of the hexafluorobenzene rings. The free-rotation of the benzene molecules cannot be represented in figure 8.5(a), due to limitations in the software.

## 8.3 Clathrates

These are solids made from water molecules that have an extended ordered structure full of voids, and small molecules that occupy those voids. They are like molecular sponges, where the large internal voids are filled with small molecules, and without these unreactive and inert included molecules the structure would not form. These materials occur naturally and the most abundant form is the clathrate of water with methane with the general formula of $CH_4$ ($5.75H_2O$).

(a)

(b)

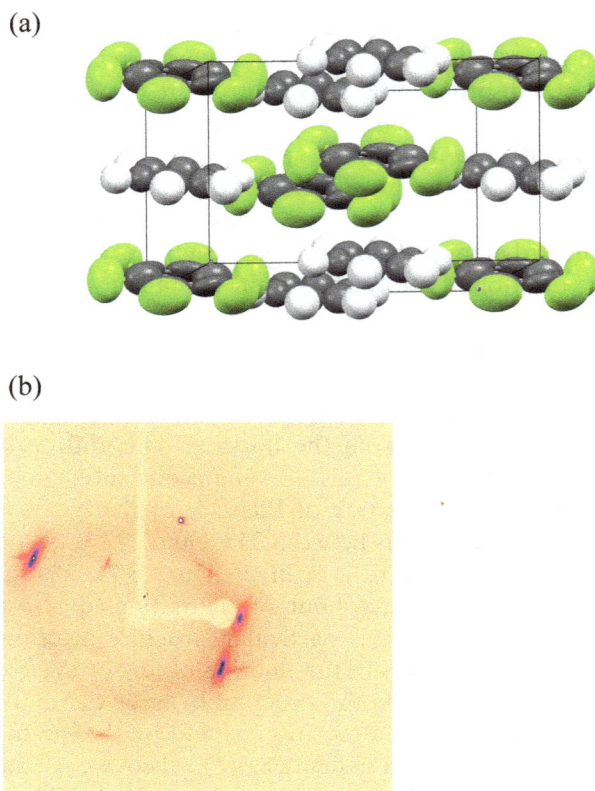

**Figure 8.5.** (a) A view of the structure of Phase II of the binary adduct benzene:hexafluorobenzene; the hydrogen atoms are white with green being used to identify the fluorine atoms. The $c$-axis of the crystal runs vertically, perpendicular to the planes of the molecules. (b) Some raw diffraction data from a small group of single crystals of Phase II of benzene:hexafluorobenzene; one can clearly see the thermal diffuse scattering arising from motion within the crystal (the diffraction data is smeared out into streaks, but with a well-defined reflection in some of those broad features). Analysis reveals that the motion responsible for this diffuse scattering is in the plane perpendicular to the $c$-axis; that is, only two (Miller) indices are involved and we may ascribe this motion to the light benzene molecules, which are perpendicular to the column axis which is parallel to the $c$-axis. The white dots in the diffuse scattering are coherent Bragg peaks.

Clathrate hydrates (or gas clathrates, or gas hydrates, or solid solutions) are actually crystalline materials, which are manifestations of the open three-dimensional, hydrogen bonded structure of water and ice. There is no chemical reaction between the water molecules of the host structure and the included molecules; they are solid solutions. The included molecules 'rattle about' inside large voids in the solid framework of water molecules. The amount of rattling they can do is determined by the available thermal energy. Consequently, given the fragile nature of the structure of these materials (they are solids which exist only because of weak van der Waals forces), they dissociate easily into host and caged molecules. Indeed, they are only found to exist at relatively extreme conditions; for example the low temperatures and high pressures of the seabed of the Continental Shelf. Here, the temperature and the pressure are such that the moment methane gas seeps from the rocks it is immediately *swallowed up* by the abundant water to form a clathrate.

The nominal methane clathrate hydrate composition is $(CH_4)_4(H_2O)_{23}$, corresponding to 13.4% methane by mass, although the actual composition is dependent on how many methane molecules fit into the various cage structures of the water lattice. The observed density is around 0.9 g cm$^{-3}$, so methane hydrate will float to the surface of the sea or of a lake unless it is bound in place by being formed in or anchored to sediment.

Methane forms a structure I hydrate with two dodecahedral (12 vertices, thus 12 water molecules) and six tetradecahedral (14 water molecules) water cages per unit cell; see figure 8.6. This crystal belongs to the cubic system and the isometric-hexoctahedral crystal class, and is characterized by a unit cell length of 11.877 Å; the occupancy (Z) is 3. Because of sharing of water molecules between cages, there are only 46 water molecules per unit cell. This compares with a hydration number of 20 for methane in aqueous solution. The brown and pink atoms in figure 8.5 represent the carbon and hydrogen of the methane molecule, which sits inside the cage and spins even at temperature well below 77K; and so the crystallographer when trying to fit the measured diffraction pattern for this material will discover that the methane molecules are disordered, or rather the carbon atom may be located but not the positions of the four hydrogen atoms.

The key point here is that the guest molecule, the methane molecule in the structure depicted in figure 8.6 cannot be located with precision using x-ray diffraction. Even at very low temperatures, the methane molecule in the large void will be poorly located and it will be spinning even at low temperatures. What this means it that there is orientational disorder in the crystals of methane-ice; there are many possible orientations of the methane molecule in the clathrate, and all

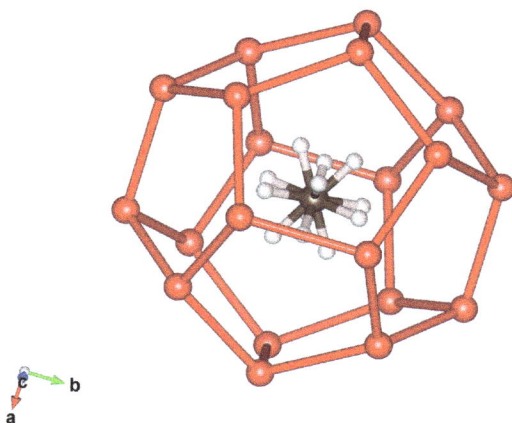

**Figure 8.6.** One of the cages in the methane hydrate I crystal structure. The red atoms in this structure are the oxygen positions in the water molecules (their hydrogens have been removed for clarity). Image from https://crystallography365.wordpress.com/2014/01/18/fire-ice-the-structure-of-methane-hydrate-i/.

these orientations are close in energy on a flat isotropic energy surface, and at low temperatures all these sites are populated. Consequently, the precision of the final fit of the x-ray data to a model of the structure will be lacking, leading to uncertainties. Not being able to fit the structure of a plastic crystal to the same levels of precision achievable in inorganic solids is not a comment on the experimenter, but is merely a reflection of the complexity of the subtle interactions in the plastic crystal and the clathrate; that is, to the absence of any particularly strong intermolecular forces.

Clathrate hosts can, however, be formed from other molecules that are capable of forming networks of hydrogen bonds, and some of these, if the generated void is large enough, can host large inert molecules. And a detailed investigation of the structure of such clathrates can tell us a good deal about the forces operating between the guest lattice and the host molecules.

Buckminsterfullerene, $C_{60}$, and hydroquinone (HQ) form a 1:3 complex $[C_6H_6O_2]_3C_{60}$. This complex was first identified in 1991 [2] as a clathrate, where the hydroquinone molecules form a 3-dimensional hydrogen-bonded host network with cavities just large enough to trap a single $C_{60}$ molecule via van der Waals interactions. The HQ host builds up a single H-bonded super-network with super-cubes as building blocks, which accommodate the large $C_{60}$ guest molecules. The enclathration of $C_{60}$ is rather tight owing, in part, to favorable host–guest electro-static interactions (a $\pi$-cloud interaction as seen in figures 2.2 and 2.3).

The diameter of the cavity in the HQ framework is 10 Å, and so just about equal to the van der Waals diameter of $C_{60}$. It was also seen that the guest $C_{60}$ molecules are orientationally disordered. This is of interest because it provides information about the details of the interaction between host and guest molecules. If the disorder is dynamic, it also permits the calculation of the average potential experienced by a molecule, as would be the case for the average potential experienced by a methane molecule in an ice clathrate such as in figure 8.6. It is seen that each $C_{60}$ molecule is coordinated by twelve hydroquinone molecules, six face-on and six edge-on, which is what one would expect from a maximizing of the asymmetries of the $\pi$-cloud of electrons on the HQ molecules with those on the $C_{60}$ molecules; a combination of quadrupole moment–quadrupole moment interactions and bond dipole moment–quadrupole moment interactions (as depicted in figures 2.2 and 2.3 for solid benzene). Thus the crystal architecture arises from a combination of weak non-bonded interactions; typical hydrogen bonds for the host structure of the clathrate, and a variety of bond dipole moment to $\pi$-cloud and $\pi$-cloud to $\pi$-cloud interactions, as seen in the binary adducts benzene:hexafluorobenzene and mesityle:hexafluorobenzene.

## Further reading

[1]  van Nes G J H and Vos A 1978 *Acta Crystalogr.* **B34** 1947–56
[2]  Ermer O 1991 *Helv. Chim. Acta.* **74** 1339–51

# Chapter 9

## Giant covalent structures: diamond and graphite

At ambient conditions, the stable bonding configuration of carbon is graphite. There is an energy difference, or activation barrier between graphite and the other common form of carbon, diamond, of order 0.4 eV per atom. However, due to this energy barrier between the two allotropes of carbon, the transition from diamond to the stabler form of graphite at normal conditions is very slow. This transition can, however, occur more rapidly when diamond is exposed to high temperature. A simple energy level diagram demonstrates why diamond (a metastable material) does not spontaneously decompose into graphite, the most energetically stable form of carbon under ambient conditions[1].

Both graphite and diamond exist at ambient temperatures and pressures, and both are composed of carbon atoms only. Yet these two covalently bonded giant structures are very different in their physical properties. A difference that is determined by the very different covalent bonds that exist between the carbon atoms.

Carbon has an electron configuration of $1s^2 2s^2 2p^2$. In diamond, each carbon shares electrons with four other carbon atoms; forming four single covalent bonds. That is, the atoms of carbon have bonded via $sp^3$-hybrid atomic orbits, as do the carbon atoms in methane and ethane. Even though metastable, diamond has a very high melting point—almost 4000 °C.

Figure 9.1 shows the arrangements of carbon atoms in diamond; each carbon atom is attached to four other carbon atoms. Diamond belongs to the cubic system and to the hexoctahedral class, and is characterized by a unit cell length of 3.57 Å. This structure can be imagined as a single giant covalent structure; that is, it is capable of filling all space and a diamond may be thought of as a single molecular entity; and some diamonds are large (the Cullinan Diamond weighed 3106.75 carats or 621.35 g). The arrangements of carbon atoms given in the above figure do not

---

[1] It's all about thermodynamics; see the graphs at http://geologylearn.blogspot.co.uk/2015/11/elementary-concepts-of-thermodynamics.html

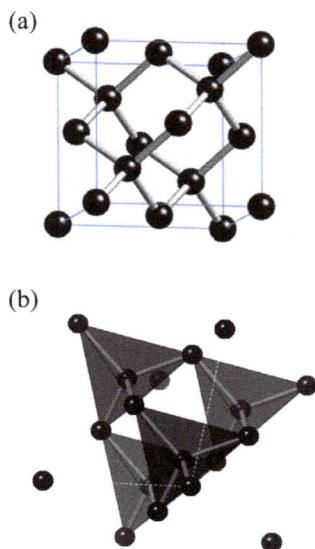

**Figure 9.1.** Two views of the structure of diamond. Part (a) shows the lattice of atoms, and part (b) permits us to see the tetrahedral disposition of bonds, which is so familiar to organic chemists. An animated image of the structure of diamond may be seen at https://www.youtube.com/watch?v=vJZL1lLco44. See http://www.chemtube3d.com/gallery/structurepages/diamond.html for an interactive animated structure.

represent a molecule; it is not a stable subunit that could come together with other similar subunits, or molecules and through non-bonding interactions give rise to a crystal, because the number of atoms in a real diamond is completely variable, depending on the size of the crystal. One must imagine that at the hard surface of a diamond crystal there will be other molecules bonded to the vacant *dangling* bond site on the outermost carbon atoms of the crystal.

Known to the Ancient Greeks as ἀδάμας—adámas (unalterable, or unbreakable), diamond is the hardest known naturally occurring material (10 on the Mohs' scale of mineral hardness). Diamond is extremely strong owing to the bonded structure of its carbon atoms, where each carbon atom has four neighbors joined to it with $sp^3$-hybrid covalent bonds. The atoms of carbon are tightly bonded together to form four covalent bonds, with the electrons of the bonded atoms shared equally between the bonded pair. Consequently, diamond is an electrical insulator. And for the same reason, it is insoluble in water and organic solvents. There are no possible attractions which could occur between solvent molecules and the bonded carbon atoms which could overcome the attractions between the covalently bound carbon atoms.

The precise tensile strength of diamond is unknown, however, strength up to 60 GPa has been observed, and it could be as high as 90–225 GPa depending on the perfection of diamond lattice and on its orientation: Tensile strength is the highest for the [100] crystal direction (normal to the cubic face), smaller for the [110] and the smallest for the [111] axis (along the longest cube diagonal). Diamond also has one of the smallest compressibilities of any material. Boron nitride when in a form structurally identical to diamond, that is, the zincblende structure[2], is nearly as hard as diamond. It has been shown that some diamond aggregates having nanometer grain size are harder and tougher than conventional large diamond crystals, thus they perform better as abrasive material.

Cubic diamonds have a perfect and easy octahedral cleavage, which means that they only have four planes; that is, weak directions following the faces of the octahedron where there are fewer bonds, along which diamond can be split upon blunt impact to leave a smooth surface. Similarly, diamond's hardness is markedly directional: the hardest direction is the diagonal on the cube face, 100 times harder than the softest direction, which is the dodecahedral plane. The octahedral plane is intermediate between the two extremes. The diamond cutting process relies heavily on this directional hardness, as without it a diamond would be nearly impossible to fashion. Cleavage also plays a helpful role, especially in large stones where the cutter wishes to remove flawed material or to produce more than one stone (large single crystal) from the same piece of rough diamond.

Graphite is another giant covalent structure; that is, a single molecule extending into macroscopic space, but the arrangement of carbon atoms is entirely different in graphite than in diamond. Graphite has a layered structure. Figure 9.2 shows the arrangement of the carbon atoms in each layer, and the manner in which the layers are stacked on top of each other. Graphite belongs to the hexagonal system and to the dihexagonal-bipyamidal crystal class; the unit cell is characterized by the following dimensions: $a = b = 2.461$ Å and $c = 6.708$ Å, and the occupancy of the unit cell is 4. The layers of carbon atoms are 3.354 Å apart. The layers can, of course, extend over large distances; one can cover a surface entirely with hexagons.

Each carbon atom in graphite uses three of the four available bonding electrons to form covalent bonds to its three closest neighbours. That leaves a fourth electron in a bonding level. These additional electrons in each carbon atom become delocalized over the whole of the sheet of atoms that constitute one particular layer. This is akin to the $sp^2$-hybrid bonding seen in ethylene, but whereas in ethylene the additional two carbon electrons form a $\pi$-bond, in graphite these electrons remain delocalized. They are no longer associated directly with any particular atom or pair of atoms, but are free to wander throughout the whole sheet. This delocalization is analogous to the situation seen in aromatic molecules such as benzene, where there is neither a double C=C bond, nor a single C–C bond, but an intermediate aromatic bond and the extra electron density is delocalized above and below the plane of the hexagonal

---

[2] The zincblende structure is named after the mineral zincblende, or sphalerite, one form of zinc sulfide ($\beta$-ZnS). As in the rock-salt structure, the two atoms form two interpenetrating face-centered cubic lattices. However, it differs from the rock-salt structure in how the two lattices are positioned relative to one another. The zincblende structure has tetrahedral coordination; each atom's nearest neighbors consist of four atoms of the opposite type, positioned like the four vertices of a regular tetrahedron. Altogether, the arrangement of atoms in zincblende structure is the same as the diamond cubic structure, there are alternating types of atoms at the different lattice sites; examples of compounds with this structure include zincblende itself, lead nitrate, and many semiconductors.

(a)

(b)

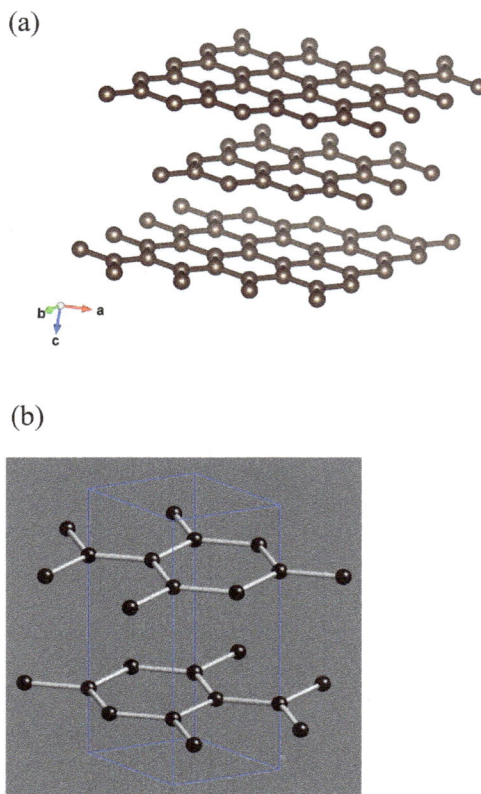

**Figure 9.2.** Two views of the structure of graphite. Part (a) gives a reconstruction of the arrangement of carbon atoms, and part (b) gives the disposition of the unit cell. See http://www.chemtube3d.com/ClaydenCarbonAllotropes.html for an animated interactive structure, and for the structures of a range of allotropes of carbon.

molecule, giving the observed large electric quadrupole moment. In graphite, such delocalization of electron density occurs over the entire layer of carbon atoms, which can have macroscopic dimensions. There is, however, no direct contact between the delocalized electrons in one sheet and those in the neighbouring sheets.

The atoms within a sheet are held together by strong covalent bonds. In fact, these covalent bonds are stronger than those in diamond, because of the additional bonding from by the delocalized electrons. The sheets of carbon atoms, however, are held together by weak, non-bonded van der Waals dispersion forces. As the delocalized electrons move around in the sheet, large dipoles can be generated for short periods of time, which will induce dipoles of opposite phase in the sheets above and below. And so, throughout the whole graphite crystal there will be fluctuating induced dipole–induced dipole attractive interactions. This is essentially the theory of London to explain the condensation of gases such as neon (see chapter 4).

Due to this unusual structure and the delocalized nature of the electrons in the solid, graphite has some contradictory properties. Graphite has a high melting point, similar to that of diamond. In order to melt graphite, it is not enough to loosen one sheet from another (the bonding energy of one sheet to another sheet is modest at 43 meV per atom or about 4 kJmole$^{-1}$). You have to break the covalent bonding throughout the whole structure. But because the forces holding the layers together are very much weaker than the forces holding the hexagons of carbon atoms together in a sheet, graphite has more interesting mechanical properties than does diamond. Whereas diamond is the hardest known substance, graphite has a soft, slippery feel, and is used in pencils and as a dry lubricant. You can think of graphite rather like a pack of cards—each card is strong, but the cards will slide over each other, or even fall off the pack altogether. When you use a pencil, sheets of carbon atoms are rubbed off and stick to the paper. Bonding between layers is via weak van der Waals bonds, which allows layers of graphite to be easily separated, or to slide past each other.

As graphite has a layered structure, it has a lower density than diamond, due to the relatively large amount of space (vacuum) that exists between the sheets. And for the same reason as diamond, graphite is insoluble in water and organic solvents. Attractions between solvent molecules and carbon atoms will never be strong enough to overcome the strong covalent bonds in graphite. But because of the delocalized electrons within the structure of graphite, the solid conducts electricity. The delocalized electrons are free to move throughout the sheets. The fourth electron is free to migrate in the plane, making graphite electrically conductive. However, it does not conduct in a direction perpendicular to the plane.

In the layered, planar structure of graphite, the individual layers are called graphene sheets. In diamond there is only one characteristic bond length, which is 1.55 Å. In graphite there are two relevant interatomic distances; namely 1.42 Å within the graphite sheets and 3.35 Å between the sheets, defining the 2-dimensional structure. Carbon nanotubes display a more complex variety of C–C bond lengths; there are four different characteristic distances: carbon–carbon bond distance within each shell (which is almost the same as in graphite), the distance between shells (which is similar to the distance between the graphite layers), the length of the tube (which can reach up to 1 mm) and the radius of the tube (which can vary from 10 Å to 100 Å). Buckminsterfullerene or $C_{60}$ has two characteristic distances; the C–C bond length which is 1.40 Å and 1.45 Å for double and single bonds, respectively.

## 9.1 The electrical properties of 2-dimensional arrays of carbon atoms

There are a number of interesting modifications to the electrical properties of ensembles of carbon atoms, which are demonstrated when going from benzene to graphite. The symmetry of the graphite crystal implies that each carbon atom, because of its $sp^2$ hybrid nature will have an associated electric quadrupole moment.

In the same way that electric and magnetic dipoles are aligned in ferroelectric and ferromagnetic crystals, molecular quadrupoles can also be oriented such that a crystal possesses a large permanent quadrupole moment. In graphite, by virtue of the electrons above and below the plane of each layer of carbon nuclei, a piece of graphite will have a macroscopic quadrupole. These macroscopic quadrupoles will give rise to an electrostatic field at the surface of the crystal, which is important in the modelling of molecules interacting with the solid surface. And in any theoretical prediction of the electrostatic interactions of graphite or graphite-like carbon surfaces (graphene, nanotubes and $C_{60}$) with other atoms and molecules, the value for the quadrupole moment of a $sp^2$-hybridised carbon atom will also be of central importance.

The direct measurement of quadrupole moments of isolated molecules is not easy, owing to the difficulties in obtaining sufficiently large uniform laboratory field-gradients. However, the field-gradient induced birefringence method can yield reliable results. It was first applied by Buckingham and Disch to $CO_2$ [1] and has subsequently been successfully used to measure the quadrupole moments of a wide range of polar and non-polar molecule [2]. A straightforward method for determining the quadrupole moment of a macroscopic crystal, based on the measurement of the time period of a known mass of quadrupolar material oscillating in a laboratory field-gradient, was briefly outlined in a review from 1959 [3], and it was successfully demonstrated with graphite in 1993 [4].

If a crystal unit cell has a centre of inversion, then the total quadrupole moment of a piece of single crystal is the sum of the moments of all the unit cells in the sample,

$$\Theta_{\alpha\beta}{}^{\text{total}} = \sum_i \Theta_{\alpha\beta}{}^i.$$

For the case of identical parallel atomic quadrupoles, as with graphite, the total quadrupole is simply

$$\Theta^{\text{total}} = N\Theta^i,$$

where $N$ is the total number of carbon atoms. A quadrupolar crystal oscillating on the end of a torsional fibre will experience a torque in the presence of an applied electric field-gradient, resulting in a change in the period of the oscillation. By measuring the dynamics of a graphite crystal in various field-gradients it has been demonstrated that it is possible to determine the quadrupole moment of the crystal and hence determine the quadrupole moment of a carbon atom in graphite due to its $sp^2$ hybridized atomic orbital [4]. The quadrupole moment of each carbon atom in a crystal of graphite has been found to be $\Theta = -3.0 \times 10^{-40}$ C m$^2$. Table 9.1 lists the quadrupole moment of some aromatic molecules determined experimentally or estimated theoretically. A simple estimate of the quadrupole moment of a carbon atom in graphite can be found by comparing the quadrupole moment per carbon atom for these compounds. From this, we would expect the quadrupole moment to be about $-4 \times 10^{-40}$ C m$^2$. This estimate is deficient for several reasons, not least because the polarization of the C–H bonds of the aromatic molecules expected from a difference in electronegativities of these two atoms will also contribute to the quadrupole moment of the molecule. Although the individual C–H bonds in these molecules have dipole moments, the vector sum of these bond dipole moments is

**Table 9.1.** Electrical properties of carbon atoms in different materials.

| System | $N_c$ | $\Theta_{zz}/10^{-40}$ C m$^2$ | $\Theta_{zz}/N_c$ ($10^{-40}$ C m$^2$ molecule$^{-1}$) |
|---|---|---|---|
| Benzene [2] | 6 | −29 | −4.8 |
| Naphthalene [5] | 10 | −46 | −4.6 |
| Anthracene [5] | 14 | −61 | −4.4 |
| Tetracene [5] | 18 | −78 | −4.4 |
| Pentacene [5] | 22 | −96 | −4.4 |
| Graphite [4] | Very large | | −3.0 |

zero for the molecule. The experimentally determined value for the quadrupole moment for carbon atoms in graphite demonstrate the polarity of the $(\delta-)$C–H$(\delta+)$ bonds in simple aromatic hydrocarbons.

Table 9.1 gives the quadrupole moments, $\Theta$, for some aromatic molecules (the $z$ axis is taken to be perpendicular to the plane containing the carbon nuclei; $N_c$ is the number of carbon atoms in the molecule; $\Theta$ is given in the SI units of $10^{-40}$ C m$^2$).

# Further reading

[1] Buckingham A D and Disch R L 1963 *Proc. R. Soc. London, Ser.* A. **273** 275
[2] Battaglia M R, Buckingham A D and Williams J H 1981 *Chem. Phys. Lett.* **78** 421
    Battaglia M R, Buckingham A D, Neumark D, Pierens R K and Williams J H 1981 *Mol. Phys.* **43** 1015
[3] Buckingham A D 1959 *Chem. Soc. Quart. Rev.* **13** 183
[4] Whitehouse D B and Buckingham A D 1993 *J. Chem. Soc. Faraday Trans.* **89** 1909–13
[5] Calvert R L and Ritchie G L D 1980 *J. Chem. Soc., Faraday Trans.* **2** 76 1249
    Califano S, Righini R and Walmsley S H 1979 *Chem. Phys. Lett.* **64** 491
    Chablo A, Cruickshank D W J, Hinchliffe A and Munn R W 1981 *Chem. Phys. Lett.* **78** 424
    Ryno S M, Lee S R, Sears J S, Risko C and Brédas J-L 2013 *J. Phys. Chem.* **117** 13853–60

# Chapter 10

# Structural elements in covalent crystals

If one allows positively- and negatively-charged spheres to interact, the electrostatic force of interaction will generate order in the system; it is inevitable. An architecture will be generated that is like that of sodium fluoride through the powerful electrostatic interaction of the electrical charges. Indeed, the spheres need not be charged to generate an architectural order; if the individual particles have a large mass and may interact through gravity; after all, stacked cannon-balls are able to display a crystalline-like order. And helium and neon atoms interact to form crystalline structures, but structures that are stable only at low temperatures.

## 10.1 Packing aromatic molecules

Given that small aromatic molecules are planar and disc-like, with highly anisotropic electromagnetic properties, the manner in which such molecules pack in solids can tell us a good deal about the nature of the forces between molecules in solids, and how crystals are engineered by self-assembly. Stacking of molecules in columns is the essential spatial arrangement of aromatic ($\pi$-cloud) systems characterized by their parallel orientation; that is, a system dominated by interacting clouds of $\pi$-electrons, with a separation between the planar surfaces of 3.45–3.7 Å, as seen in the many thousands of structures contained in the Cambridge Structural Database, and in figures 2.3 and 2.4. However, such planar molecules, when formed into crystals do not form simple stacks as would be seen in a stack of identical plates, as this would not be a stable structure; as is demonstrated by the three phase-transitions seen in solid benzene:hexafluorobenzene, and the absence of phase transitions in pure benzene and in pure hexafluorobenzene. To engineer a crystal, one also has to take into consideration the interactions between the columns of flat disc-like molecules. A crystal cannot be truly 1-dimensional. As it happens, the best known example of such stacking is not found in solid benzene, but in the parallel steps of the nucleoside bases that form the DNA molecule. The spacing between the nucleotide bases in DNA is 3.4 Å, which is in good agreement with the spacing of, for example,

3.49 Å and 3.63 Å seen in the binary adduct mesitylene:hexafluorobenzene. In DNA, the only orientation of the planar aromatic nucleoside bases is the near, or slipped parallel arrangement depicted in figure 2.2(b). Because the DNA molecules is a double helix, the nucleoside bases are naturally in a slipped parallel ordering.

We saw in chapter 2, how simple models of intermolecular electrostatics may be used to rationalize the architecture of solids such as benzene and the adduct benzene: hexafluorobenzene; to explain the origin of the slipped parallel and T-shaped arrangements of molecular quadrupole moments (the herringbone structure). Consequently, aromatic hydrocarbons are convenient models for analysis and prediction of their crystal structures and intermolecular interactions due to their well-defined and fixed molecular shapes (there is no conformational freedom). Equation (2.5) allows us to estimate the strength of the interactions within a crystal where the intermolecular spacing is $R$; for example, the attractive quadrupole–quadrupole interaction term varies as $R^{-5}$ and so given that the quadrupole moment of benzene is $-29 \times 10^{-40}$ C m$^2$ and that the flat molecules are separated by, typically, 3.5 Å, the attractive force between the molecules in a column of benzene molecules would be about 6.6 kJ mole$^{-1}$. This is about one third the strength of a strong hydrogen bond, as in ice; yet is significantly larger that the energy available at ambient temperatures ($k_\mathrm{B}T$ at 300 K = 2.5 kJ mole$^{-1}$), but this closeness to the energy available in thermal excitations does explain why such solids are susceptible to phase transitions, and why they melt near 300 K. In the case of the nucleoside bases in DNA, as these molecules are more akin to naphthalene than benzene, and as the quadrupole moment of naphthalene is larger than that of benzene ($-46 \times 10^{-40}$ C m$^2$), this attractive interaction will be larger. Such interactions in DNA are essentially $\pi$–$\pi$ electron interactions, proportional to the area of $\pi$-overlap of the molecules.

However, in addition to looking at the strength of the interactions along a column of planar molecules in a covalent solid, one needs also to consider the interactions between the columns; that is perpendicular to the column axis. It has been demonstrated [1] that packing arrangements of aromatic hydrocarbons depend on the number and position of C and H atoms in the interacting molecules. Besides an obvious geometrical tendency of aromatic systems to pack in a near parallel manner (a strong attractive interaction and the efficiency of filling space), there is an additional, strongly orientationally dependent interaction that generates crystal architectures. These are bond dipole–bond dipole interactions; that is, C–H$\cdots\pi$-cloud, and C–H$\cdots$H–C interactions. These are Coulombic-like interactions, which depend on net charge distribution and are responsible for the T-shape arrangements of aromatic molecules where the C–H bond dipole on one ring will be in the appropriate orientation to point towards the delocalized $\pi$-cloud of a neighbouring molecule (see figure 2.2(c) and the structure of benzene in figure 2.3).

Replacement of a C–H group by a N atom in the aromatic six-membered aromatic ring significantly alters the charge distribution, and so changes the stacking parameters. The presence of a N atom in an aromatic ring changes both the $\pi$- and $\sigma$-electrons distributions. The electric quadrupole moment of hexafluorobenzene is $31.7 \times 10^{-40}$ C m$^2$, but that of $s$-triazine is about a tenth of this value, and for the same reasons that the quadrupole moment of 1,3,5-trifluorobenzene is the arithmetic mean of the quadrupole moments of benzene and hexafluorobenzene.

## 10.2 Interacting bond dipole moments

Hydrogen bonding is often cited as being the clearest example of molecular recognition, as it can only occur between certain atoms separated by specific distances. And because of these limits, inserting those particular atoms in a designed molecular structure is a sure way of inserting hydrogen bonds, and the stability generated by such bonds into a crystal. That is, locating and inserting hydrogen bonds is a reliable design element in crystal engineering. Consequently, hydrogen bonds of the RO–H···O=CR, RO–H···O–CR, RN–H···O=CR, RN–H···O–CR, RN–H···O–H, and RO–H···N–H type have been extensively studied and employed in the designed synthesis of bespoke supra-molecular assemblies. In contrast, less is known about hydrogen bonds or hydrogen bond-like links of the RO–H···F–CR and RN–H···F–CR type, or of the RC–H···F–CR type.

The weaker hydrogen bonds of the later type, RC–H···F–CR, have also been suggested as a means of stabilizing crystal structures [1] as the C–H bond is known to be a hydrogen-bond donor. While it has been observed that C–H···F–C interactions are weak, they do make a contribution to crystal architectures. Yet, it must be admitted that there is a continuous scale of strength between strong hydrogen bonds, where a hydrogen atom is shared equally by two electronegative atoms, and a non-bonded interaction; that is, when is a van der Waals interaction a hydrogen bond-like interaction and when is it a dipole–dipole interaction? This raises the question as to whether or not, non-bonding intermolecular interactions weaker than a classical hydrogen bond (of order, 22–23 kJ mole$^{-1}$) become part of a useful design procedure in bespoke supramolecular syntheses; that is, a useful fusing element in crystal architectures?

These weak intermolecular interactions are perhaps better described as bond dipole–bond dipole interactions or quadrupole moment–bond dipole moment interactions. These are weak electrostatic interactions, which in crystals of organic molecules do play a significant part in determining the structure of the solid, and in defining the dynamics of the crystal architecture; for example, in initiating the various sold state phase transitions seen in binary adducts such as benzene: hexafluorobenzene and mesitylene:hexafluorobenzene, which involve interactions between RC–F bonds on a molecule in one of the close-packed columns that constitute the crystal and RC–H bonds on a molecule in an adjacent column. Thereby giving stability in the direction perpendicular to the axis of the closely packed columns. Such lateral interactions are sometimes termed hydrogen bonds, but in fact, they are simple dipole–dipole interactions; however, as the molecules have no permanent electric dipole moments, we must consider the asymmetry of the charge in the bonds of the molecules. In benzene and hexafluorobenzene, for example, the individual C–H and C–F bonds are polarized (the electronegativities of H, C and F are 2.3, 2.5 and 4.2, respectively), and the vector sum of the six bond dipoles is zero in both molecules, but the bond will possess bond-dipole moments.

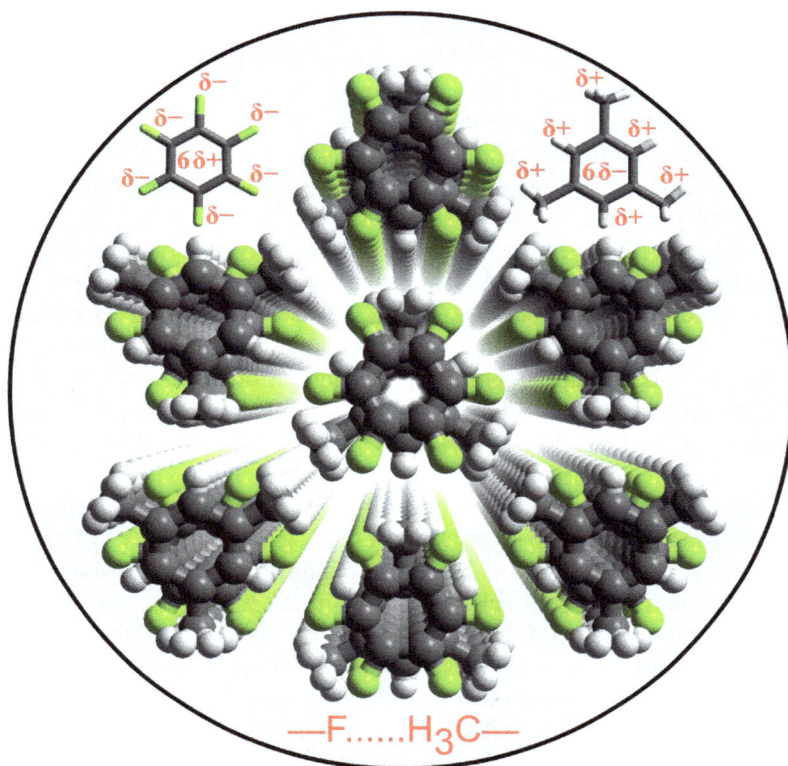

**Figure 10.1** Structure of the lowest temperature phase (III) of the binary adduct, mesitylene:hexafluoroben-zene, looking down the *c*-axis of the crystal; the carbon atoms are black, the hydrogen atoms are white and the fluorine atoms are green. The small figures in red are to highlight the very different, but related, intermolecular bonding seen in this material: there is the quadrupole–quadrupole interaction that generates the columns and then the dipole–dipole and dipole–quadrupole interactions that hold the columns together.

Figure 10.1 demonstrates the complex interactions of bond dipole moments in organic crystals. We see the structure of the lowest temperature phase (III) of the binary adduct mesitylene:hexafluorobenzene. The structure consists of close-packed columns of alternating mesitylene and hexafluorobenzene molecules. The columns are held together by the attractive interactions between the quadrupole moments of the constituent molecules, and the image shows the close interactions of the C–H and F–C bonds in one column with those bonds in neighbouring columns; that is, there is a network of interacting bond dipole moments. The bond dipole–bond dipole interaction between neighbouring columns (C–F⋯H–C) provides a localized attractive potential perpendicular to the *c*-axis of the crystal, which is parallel to the column axis. The small red schematic figures in figure 10.1 summarize the inter- and intracolumnar attractive electromagnetic forces (quadrupole–quadrupole interactions within a column and dipole–dipole interactions between columns). The structure of the lowest temperature phase (IV) of the binary adduct, benzene:hexafluorobenzene, seen in figure 2.4 (again looking down the *c*-axis of the crystal) also shows the interpenetrating C–F and C–H bonds on adjacent columns—like the cogs of the wheels in a gear-box. The structure again comprises closed-packed columns of quadrupolar bound molecules, and these columns are bound to each other by bond dipole–bond dipole interactions.

Estimating the magnitude of these intercolumnar bond dipole–bond dipole interactions ($E_{DD}$) is difficult as one must assume a magnitude for the bond dipole moment. Crystallographers [1] have shown that the C–H$\cdots$F–C distances that are most important to defining the structure of a range of crystals of partially fluorinated benzene are in the range 2.45–2.8 Å. If we take this range and use equation (2.5), we find that for the shortest distance, $E_{DD} = 4.2$ kJ mole$^{-1}$, and is 3 kJ mole$^{-1}$ for the upper limit of the measured distances. Whether such interactions should be termed hydrogen bonds or bond dipole–bond dipole interactions is a moot point given the continuum of strengths of non-bonding interactions. One can see therefore how simple models of molecular electrostatics, or molecular mechanics allow us to explain the stability of a range of organic crystals.

The presence of such hydrogen bond-like interactions has a major role in generating stability in many crystals. Figure 10.2 gives two views of the structure of (RS)-N-methyl-1-phenylpropan-2-aminium chloride (the hydrochloride salt of methylamphetamine). This salt melts at about 170 °C, whereas the free base melts at a lower temperature. The crystal structure is monoclinic (the space group is P2(1), which as pointed out in chapter 3 is the space group that contains the greatest number of low-symmetry or optically active organic molecules), with unit cell parameters: $a = 7.1022$ Å, $b = 7.2949$ Å and $c = 10.8121$ Å, $\beta = 97.293°$; there are two molecules per unit cell. The crystal structure of this salt has been stabilized by the presence of hydrogen bonds, introduced into the crystal by the formation of the hydrochloride salt; these hydrogen bonds are not present in the base and are clearly seen in figure 10.2(b). Figure 10.2(a) shows the close approach of the methyl group (–CH$_3$) on one molecule to the $\pi$-cloud of the benzene rings of neighbouring molecules. This close approach will generate a bond dipole–quadrupole moment interaction that will also create stability in the crystal, but which are weaker interactions than the hydrogen bonds seen in figure 10.2(b).

An examination of equation (2.5) and the images in figures 2.3, 2.4, 10.1 and 10.2, reveal that many of the interactions that are observed between small planar molecules in organic crystals may be interpreted as bond dipole–quadrupole moment interaction, that is, the attractive electrostatic interaction is $(3/2)\,\mu_1\Theta_2$ R$^{-4}$ and/or $(3/2)\,\mu_2\Theta_1$ R$^{-4}$ (there $\mu$ and $\Theta$ are the interacting bond dipole moments and molecular quadrupole moments).

## 10.3 Vibrational dynamics in organic crystals

How does one measure details of the attractive intermolecular forces responsible for generating solids? X-ray crystallography gives a picture of the consequence of the interaction of those forces, but no information about the magnitude of those forces.

In the harmonic approximation, the vibrational modes of a polyatomic molecule are not coupled and the various vibrational quantum levels are equally spaced, so

(a)

(b)

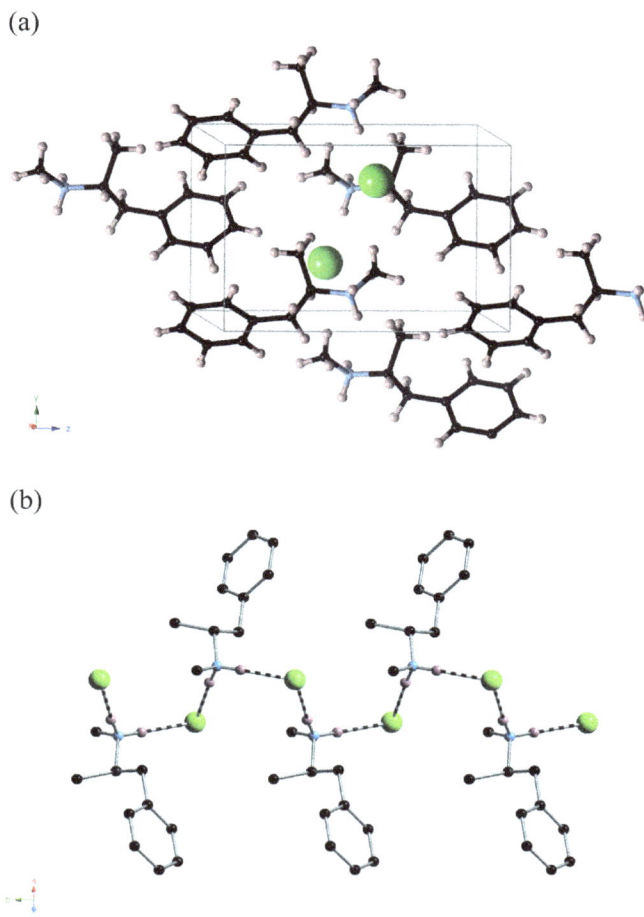

**Figure 10.2** Two views of the crystal structure of *N*-methyl-l-phenylpropan-2-aminium chloride (it is a secondary amide salt). The chlorine atoms are shown in green, carbon atoms in black, hydrogen in white and nitrogen in blue. In the crystal structure, intermolecular N–H···Cl hydrogen bonds form a network of hydrogen bonds, which stabilizes the structure, as may be seen in (b) (where hydrogen atoms not involved in the hydrogen bonding have been omitted).

hot bands where the lower level of the transition is actually above the ground state are not distinguishable from so-called fundamental transitions, which begin at the vibrational ground state. However, vibrations of real molecules always have some anharmonic character which permits coupling between the different vibrational modes and leads to change in the form of the potential energy surface upon which the molecule is vibrating.

If the vibrating molecules under consideration are organized into a solid lattice then the vibrational distortions of the covalently bonded arrangements of atoms in the molecules (bond lengths and bond angles) will be perturbed by neighbouring molecules. The solid itself will have vibrational modes, termed phonons. Indeed, the phonons seen in a crystal are closely linked to the crystal structure of that solid, and to the vibrations of the molecules that constitute the solid.

A phonon is a description of a vibrational motion in which a lattice of atoms or molecules oscillates coherently at a single frequency; they can only exist in solids because of the rigid lattice. Such coherent oscillations of the lattice are considered normal modes, because any arbitrary lattice vibration can be considered to be a superposition of the elementary vibrations of each element in the lattice. While normal modes are wave-like phenomena in classical mechanics, phonons can be thought of as having particle-like properties and play a major role in many of the physical properties of condensed matter; for example, thermal and electrical conductivity.

Solids with more than one atom in the unit cell, exhibit two types of phonons: acoustic phonons and optical phonons. Acoustic phonons are coherent movements of atoms out of their equilibrium positions. If the displacement is in the direction of propagation, then in some areas the atoms will become closer and in other areas farther apart as in a sound wave in air (hence the name acoustic). Optical phonons are out-of-phase movements of the atoms, one atom moving to the left, and its neighbour to the right. They are designated optical because in ionic crystals they are excited by infrared radiation. If $N$ is the number of pairs of atoms in a 1-dimensional representation of a crystal, then for $N \geqslant 2$, the solid will display three acoustic phonons; one longitudinal and two transverse modes. The number of possible optic phonon modes is given by the relationship $3N - 3$.

These phonon modes arise from the relative motions of one atom against another atom within the solid; that is, they represent the vibrational dynamics of the non-bonded components of two (or more) molecules held together by an attractive electromagnetic force. As these vibrations or oscillations are taking place in a relatively shallow potential[1], the frequencies of the transitions within this bounded potential are small, as can be seen in the inelastic neutron scattering experiments and the Raman scattering experimental results given in figure 10.3.

In vibrational spectroscopy, combination bands are transitions that involve a change in the quantum number of more than one normal mode. Such transitions are forbidden by harmonic oscillator selection rules, but are observed in vibrational spectra of real systems due to anharmonic couplings of normal modes. A combination band is a two-photon process; for example, the excitation $<101| \leftarrow <000|$ in a system of three modes of vibration. The frequency of a combination band is slightly less than the sum of the frequencies of the fundamentals, again due to anharmonic shifts in both vibrations.

---

[1] The model of a 1-dimensional solid usually used to explain the origin of phonons is also a representation of the solid adduct benzene:hexafluorobenzene, where the two atoms in the unit cell are in fact a benzene molecule and a heavier hexafluorobenzene molecule. In this case, because of the large difference in the mass of the two molecules, the (lower frequency) acoustic phonons represent mainly the motion of the heavier hexafluorobenzene molecule, and the (higher frequency) optic phonons represent mainly motion of the lighter benzene molecules. The attractive potential within which these two molecules are vibrating may be approximated by the interaction of the electric quadrupole moments of the two molecules or by measurements of the thermal expansion of the solid (see chapter 11.2).

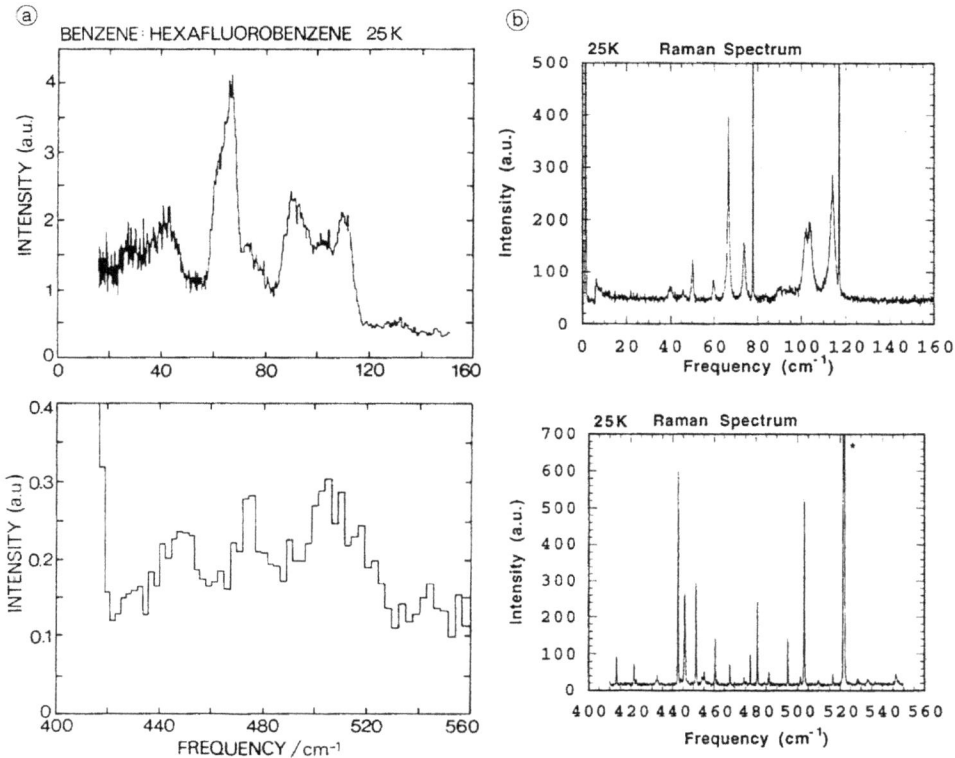

**Figure 10.3** The phonon spectra in benzene:hexafluorobenzene at 25 K determined by: (a) inelastic neutron scattering, and (b) Raman spectroscopy. In (a) upper image, we see the neutron phonon spectrum (often termed the density of states) at low frequency (up to 160 cm$^{-1}$), and in (a) lower we see the side-band or combination mode generated by the simultaneous excitation of the phonon modes with the out-of-plane vibrational mode ($E_{2u}$) of the benzene molecule, which is Raman inactive and occurs at 409 cm$^{-1}$. In (b), we see the same frequency regions, but in a Raman spectrum of significantly higher frequency resolution. Again, the combination mode of the phonon spectrum and the vibrations of the molecules is clearly seen. There are some artefacts of the laser seen in the Raman spectrum (such as the one labelled at 522 cm$^{-1}$). Images from [2].

Consequently, measurement of combination bands in solids where one is simultaneously exciting a high frequency vibrational mode in a constituent molecule and an intermolecular mode, or the phonon spectrum will provide information about the anharmonic character of the potential energy surface within the crystal; this is what is seen in figure 10.3. In the upper two images, we see the phonon spectrum of benzene:hexafluorobenzene as measured by neutron scattering (a) and Raman spectroscopy (b) at a much higher frequency resolution. The observed transitions represent the motion (translational and vibrational) of the benzene and hexafluorobenzene molecules against each other within the attractive potential that defines the crystal (see [2] for a full assignment of each of these transitions). In the lower two images in figure 10.3, we see the combination band of the phonon transitions with the out-of-plane vibration of the benzene molecule (symmetry $E_{2u}$); this vibration is Raman inactive and so does not appear in the lower image of figure 10.3(b)[2] and we see only those phonon modes that have the appropriate symmetry to appear when coupled with the molecular vibration.

In the case of the combination band between the $E_{2u}$ vibration of a benzene molecule in benzene:hexafluorobenzene and the intermolecular phonon modes, it was seen figure 10.3(b) that the phonon lines are red-shifted by as much as 20 cm$^{-1}$ compared to their position when not coupled to the distortion of the benzene molecule. These anharmonic corrections are huge given the frequencies involved, and reveal the shallow nature of the attractive intermolecular potential.

Such spectra provide a means of probing the dynamics of the molecules in the solid, and hence of probing directly the electromagnetic field that determines the crystal architecture. In the case of benzene:hexafluorobenzene, approximated to a 1-dimensional system, it is possible to follow the temperature dependence of the transverse and longitudinal lattice modes of the constituent molecules. As the crystal consists of close-packed columns of chains of alternation benzene and hexafluorobenzene molecules, the transverse modes correspond to vibrations within the potential between the columns, and the longitudinal modes correspond to transitions within the potential that exists between the molecules along a column. It is seen [2], that in benzene:hexafluorobenzene the two transverse modes move to lower frequency as the sample temperature increases, becoming equal at a temperature just below the first of a series of phase-transitions seen in this material at 205 K. At this temperature, the similar magnitudes of the frequencies of the two transverse vibrational modes tells us that the average intermolecular potential (between the close-packed columns), weighted by the mean amplitude of molecular oscillation, are equal. At these temperatures the potential has become so flat that the molecules are now able to undergo 6-fold rotation–vibration transitions within the longitudinal potential; that is, the forces holding the columns in a close-packed structure are weakening, and the amplitude of the motion of the constituent molecules relative to adjacent columns triggers a phase transition.

This is a great deal of information; information that is essential in understanding how the molecules are moving in the solid and why there are phase transitions, that is, information essential for designing new crystals with specific properties. Given that benzene:hexafluorobenzene is only one of a large class of binary adducts, most of which appear to have, at least, one phase transition below their melting point, and many of which have important applications in organic electronics and in the design of new drugs and pharmaceuticals, high-resolution solid-state vibrational spectroscopy, especially of the phonon side-bands or combination bands would appear to be a useful technology for the study of these materials.

---

[2] Inelastic neutron scattering does not possess electric dipole selection rules, so all the vibrations of a centrosymmetric vibrating system will be observable; the vibration of the benzene molecule is seen clearly in the lower left-hand panel of figure 10.3(a).

## 10.4 Why crystals melt

Boiling is the transition from a poorly ordered system (the liquid) to an un-ordered system (a gas). The other phase transition with which we are all familiar, melting, is perhaps more difficult to envisage. A solid being subject to heat behaves much like a liquid being heated. As the available thermal energy increases, the localized motions executed by the molecules fixed in their lattice positions increase in magnitude. Eventually, the available thermal motion will so weaken the intermolecular forces holding the lattice together, that the lattice will disintegrate. In other words, when the localized vibrations (librations) become so large as to disrupt the localized intermolecular bonding (when the molecules' mean kinetic energy is greater than the attractive intermolecular potentials), the ordered system transforms into a poorly ordered system; that is, it melts or it boils.

When the solid melts, the atoms and the molecules remain but the form and order that existed in the solid structure are lost. Thus, molten calcite would not be birefringent, and in molten benzene:hexafluorobenzene the pairing of the benzene and hexafluorobenzene molecules (individual dimers) would remain, but probably not at the separation seen in the solid.

One of the earliest theories of melting is due to Frederick Alexander Lindemann (1886–1957). He conjectured that for all solids, due to thermal fluctuations melting occurred at a temperature when the average displacement of the atom or molecule, which constitute the solid reached a given fraction (say, about 10%) of the average intermolecular or interatomic spacing. This somewhat vague, almost qualitative approach is a heuristic attempt to model a complex set of observations.

The amplitude of motion of atoms and molecules in solids is measured as a consequence of the determination of the crystal structure of the solid. That is, one is able to determine the relative positions of all of the atoms in a solid via x-ray or neutron diffraction and, in addition, one is often able to directly determine the amplitude of motion of the constituent atoms at the temperature of the crystal structure determination, because these amplitudes influence the precision of the final measurements.

Neutron scattering measurements have been made of the temperature dependence of the magnitude of solid state vibrations in solid benzene and in benzene: hexafluorobenzene. It was observed that near the melting temperature, the benzene molecules in solid benzene and in solid benzene:hexafluorobenzene are rotating about an axis passing through the centre of the molecule's hexagon of carbon atoms to such a degree that they are executing large jumps; a jump of 60° would cause the molecule to move to an equivalent position. As a result of these observations, it was proposed [3] that melting occurs in the solids formed from such small aromatic molecules when this mean amplitude of oscillations reaches some fraction of the free rotation amplitude, for example, some fixed fraction of 60° for benzene; an analogy with the Lindemann theory of melting.

This model of melting tells us that melting of the solid composed of small aromatic molecules occurs when the constituent molecule, or a side-group attached to the main molecule, is no longer able to store the increasing amounts of available thermal energy in the manifold of excitations associated with the low-frequency librations or rotations of the molecules within their lattice positions; for example, rotation of a methyl group, or of a benzene ring. As a consequence, this energy goes into other degrees of freedom, for example, vibrations of the weak intermolecular bonds thus destabilizing the lattice. The available thermal energy is rapidly transferred from the vibrating, or rocking molecules to the weak intermolecular bonds, leading to the break-up of the intermolecular bonding and hence to melting. Such a model is able to rationalize the order of melting temperatures in a series of molecules, but not the values of the temperatures themselves; it is merely another heuristic model, not a formal theory.

Whereas in solid benzene the molecules are able to store thermal energy by executing large-amplitude motions even within the solid and yet remain a solid; once a side-group has been introduced onto the benzene ring as in, for example, toluene ($C_6H_5$–$CH_3$) or ethylbenzene ($C_6H_5$–$CH_2$–$CH_3$), there is a stronger coupling to the rocking motion of neighbouring molecules (the methyl side-group is particularly good at facilitating the transfer of thermal energy from one part of the crystal lattice to a neighbouring part of the lattice, because it approaches close to neighbouring molecules leading to bond dipole–quadrupole interactions). Then if this side-group is unable to store the increasing levels of thermal energy in its motions, this energy must lead to disruption of the weak intermolecular bonds, and thus to lattice instability and to melting. Observations reveal that toluene melts at about the temperature where one observes the onset of thermally activated 6-fold librations of the benzene molecules in benzene and in benzene:hexafluorobenzene.

Figure 10.4 displays the structure of solid toluene, and figure 10.5 the structure of solid ethylbenzene. Toluene belongs to the monoclinic system and to the low-symmetry space group, P2(1)c, and has the following unit cell dimensions: $a =$ 7.666 Å, $b = 5.832$ Å, $c = 26.98$ Å, with $\beta = 105.73°$ with an occupancy of eight molecules per unit cell. We also see in figure 10.4(b) the herringbone structure that indicates that the crystal architecture of toluene is dominated by the interactions of the quadrupole moments of the constituent molecules (as in benzene, $N_2$, ethylene, acetylene, etc). This is not surprising as the quadrupole moment of toluene will be of the same order of magnitude as the quadrupole moment of benzene, but it does then pose a question: why does benzene melt at 5.5 °C and toluene and ethylbenzene both melt at −95 °C? Similar questions arise with ethylbenzene seen in figure 10.5. This material also belongs to the monoclinic system and in the low-symmetry P2(1)n space group, and has the following unit cell dimensions: $a = 5.6138$ Å, $b = 14.970$ Å, $c = 8.0481$ Å, with $\beta = 102.18°$ with an occupancy of four molecules per unit cell. Figure 10.5(a) again displays the classic herringbone structure of molecular interactions dominated by quadrupole moment–quadrupole moment interactions.

What is clearly seen in the structures of both toluene and ethylbenzene is the intimate contact between the C–H bond dipoles of the methyl and of the ethyl side-groups on one molecule with the $\pi$-cloud of electrons on neighbouring molecules. This closeness is particularly well seen in the structure of ethylbenzene (figure 10.5),

(a)

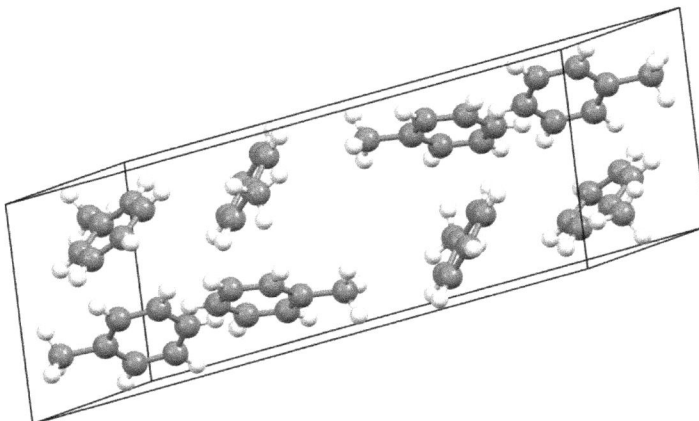

(b)

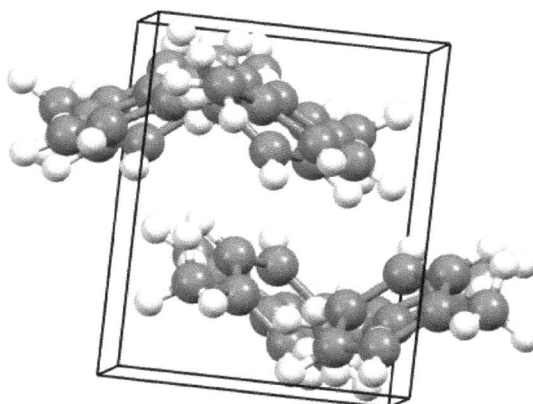

**Figure 10.4** Two views of the structure of toluene; the carbon atoms are dark grey and the hydrogen atoms are white.

where the $-CH_2-CH_3$ side-group is like the upright, curved tail of a scorpion, and is clearly a strong intermolecular link in the solid.

Investigations of the dynamics of motion in solid toluene and solid ethylbenzene [3] reveal that at temperatures just below their melting point (they both melt at −95 °C), thermally driven motion is present only in the methyl and ethyl side-groups. Because of these largish side-groups, the molecules are unable to undergo benzene-like 6-fold motion to absorb increasing amounts of thermal energy as the sample temperature rises, and as a consequence they melt at the temperature where the more symmetric benzene molecules merely start to oscillate and rotate within the solid.

(a)

(b)

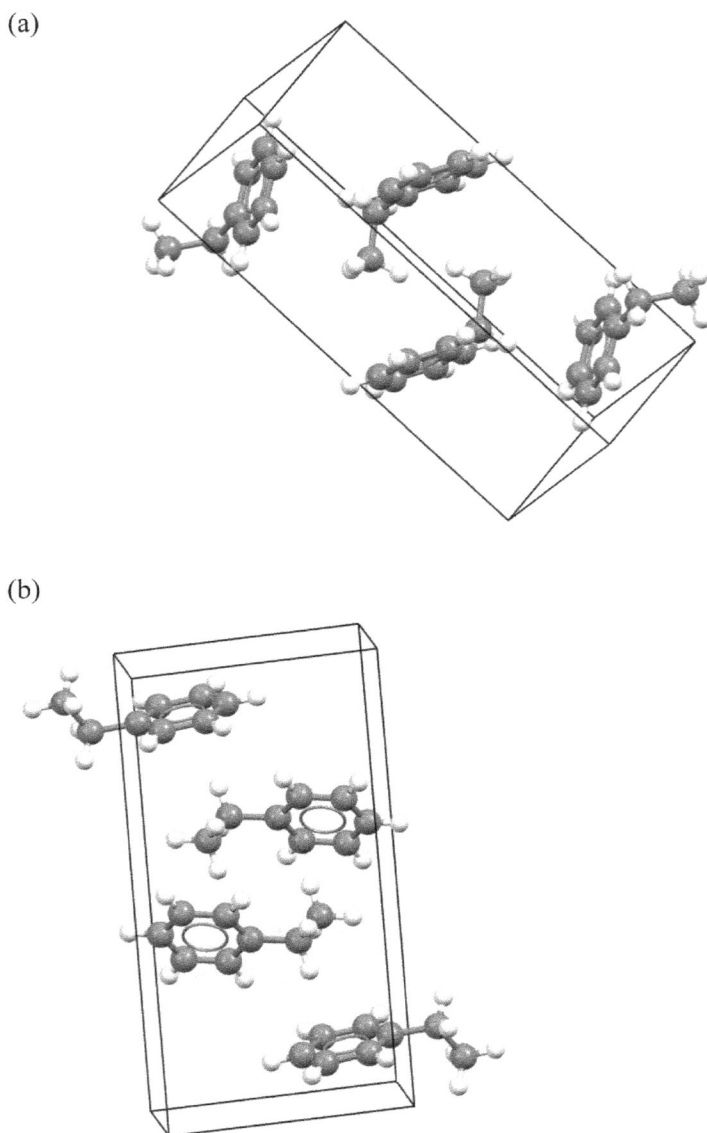

**Figure 10.5** Two views of the structure of ethylbenzene; the carbon atoms are grey and the hydrogen atoms are white.

## Further reading

[1] Desiraju G R and Gavezzzotti A 2989 *Acta Crystalogr.* B **45** 473
    Thalladi V R, Weiss H-C, Bläser D, Boese R, Nangia A and Desiraju G R 1998 *J. Am. Chem. Soc.* **120** 8702–10
[2] Williams J H and Becucci M 1993 *Chem. Phys.* **177** 191–202
[3] Williams J H and Frick B 1992 *Chem. Phys.* **166** 425–39
    Frick B and Williams J H 1992 *Europhys. Lett.* **20** 493–8

# Chapter 11

## Solids formed from aromatic molecules

The ability to predict the solid-state packing of molecules; that is, to engineer bespoke crystals, and to comprehend the observed molecular dynamics from knowledge of the electrical properties of the isolated molecules, is a goal that is much sought after. Although the strength of the various intermolecular interactions may be approximated, the problem is not straightforward. In this chapter we will look at structure–property relationships in solids formed from small aromatic molecules and we will see that crystal engineering is becoming a science rather than an art.

### 11.1 Benzene and benzene:hexafluorobenzene

As pointed out in chapter 2, the charge distribution of the simple aromatic molecules benzene and hexafluorobenzene are of interest in modelling the architecture of solids. Simple mixing of equimolar quantities of these two fluids at room temperature, produces a solid which melts at 25 °C; that is, at a higher temperature than the two pure materials. This molecular adduct arises because of the strong electrostatic interactions between the benzene and hexafluorobenzene molecules that constitute the solid in equal quantities.

Gas-phase measurements of the benzene:hexafluorobenzene dimer via molecular beam electric resonance spectroscopy, reveal that the dimer has an electric dipole moment of 0.44 Debye. This induced polarization in the dimer pair arises from the electric field of the benzene molecule polarizing the hexafluorobenzene molecule through its molecular polarizability and vice versa. With such a strong intermolecular polarization or charge transfer, the existence of a new material, a new solid with properties different from either benzene and hexafluorobenzene is perhaps not unexpected.

Benzene is the simplest member of an enormous class of chemicals. It consists of six carbon atoms in a hexagonal arrangement, and bonded to each of these carbon atoms there is a hydrogen atom with a C–C–H bond angle of 120°; the molecule is a

doi:10.1088/978-1-6817-4625-8ch11

flat hexagon, which has a radius of about 2.5 Å. Hexafluorobenzene is the same as benzene except that all the hydrogen atoms have been replaced by fluorine atoms, which gives this flat hexagonal molecule a slightly larger diameter than benzene.

For many years, benzene had been something of a crystallographic puzzle. Though it had been known since the 1860s that benzene was made up of a ring of carbon atoms, in the early years of x-ray diffraction one question that was sought by a number of researchers was the shape of the benzene ring. There was considerable debate, for instance W L Bragg thought the molecule was bent. And it was the work of E G Cox in 1932 that showed that the molecule within solid benzene was not bent and was, in fact, a flat hexagon. However, it was Kathleen Lonsdale who had paved the way for the understanding of the shape and the bonding of the benzene molecule, with her structure of hexamethylbenzene in 1929. These early studies opened the way for the field of molecular crystallography that has had a huge impact on modern solid state science and molecular biology.

Solid benzene has a structure that puts it in the orthorhombic system with the following cell parameters (at 78 K): $a = 7.292$ Å, $b = 6.742$ Å and $c = 9.471$ Å; $Z = 4$. There are no phase transition in solid benzene, and at 270 K the structure has expanded to $a = 7.460$ Å, $b = 7.034$ and $c = 9.666$ Å. The unit cell is centrosymmetric [1]. X-ray analysis of hexafluorobenzene at 120 K has shown that the space group is P2(1)/n, monoclinic with six molecules per unit cell, and $a = 16.82$ Å, $b = 9.17$ Å, $c = 5.76$ Å, $\beta = 95.8°$. Four of the molecules in the unit cell are in general positions (Wyckoff position e) and two are in special positions at centres of symmetry (Wyckoff position c). At both types of site, the molecules are planar to within 0.01 Å; in them C–C $= 1.36 \pm 0.02$ Å and C–F $= 1.32 \pm 0.017$ Å [2].

In the bulk liquid form, both benzene and its fluorinated analogue have similar physical properties, but the molecules have different electrical properties. Benzene melts at 5.5 °C and boils at 80 °C, while hexafluorobenzene melts at 5 °C and boils at 81 °C, and the two molecules are the same shape and essentially the same size. However, the manner in which the electrons are organized in the two molecules is totally different. In benzene, the electrons are essentially above and below the plane of the hexagon (of $sp^2$-hybridized carbon atoms), whilst in the fluorinated analogue, the electrons are in the plane of the hexagon, because of the greater electron density arising from the electronegativity of the fluorine atoms compared to the hydrogen atoms. The quadrupole moment of 1,3,5-trifluorobenzene is the arithmetic mean of the values for benzene and hexafluorobenzene; demonstrating the additivity of molecular properties and the utility of models of molecular quadrupole moments as partial charges and bond dipole moments.

The large negative quadrupole moment of benzene can be interpreted with the familiar picture of a delocalized cloud of electrons (electron density) above and below the plane containing the six carbon atoms. Thus, the benzene molecule is only susceptible to attack by positively-charged chemical entities (electrophilic attack) approaching the hexagonal molecule from above or below. In hexafluorobenzene, on the other hand, the electron density is now, due to the electronegativity of the fluorine atoms, contained in the plane of the carbon ring and the sign of the quadrupole moment changes, making the molecule less susceptible to chemical, electrophilic attack; an observation made and retained by every generation of organic chemistry students since the mid-19th century.

Although benzene and hexafluorobenzene have almost identical physical properties and crystallize into closely related structures, when they are mixed, a solid adduct is formed that has a melting point of 25 °C. Thus the solid, benzene: hexafluorobenzene is more stable than either of its two constituents (the solid benzene:hexafluorobenzene persists to a higher temperature than either solid benzene or solid hexafluorobenzene) and has a crystalline structure where the internal arrangements of the molecules are totally different from how benzene or hexafluorobenzene molecules arrange themselves in their respective pure solids (see chapter 2).

The difference in the interaction energy of the possible arrangements of quadrupolar molecules as seen in benzene, or nitrogen is small and is proportional to the angle between the main rotational axes of the two interacting molecules in the plane containing the two molecules represented in figure 2.1. These are the two orientations that occur in the structure if solid benzene, see figures 2.2, 2.3 and 11.1.

The structure of the lowest temperature phase (IV) of the adduct benzene: hexafluorobenzene is shown in figures 2.4 and 11.2. We see that the benzene and hexafluorobenzene molecules have become paired parallel to each other, by virtue of the strong, attraction generated by the quadrupole moments of opposite phase (the $\delta-$ clouds above and below the plane of the benzene molecules interact with the $\delta+$ clouds above and below the planes of the hexafluorobenzene molecule). The alternating sequence of equi-spaced benzene and hexafluorobenzene molecules along the crystal c-axis is clearly seen in figures 2.4 and 11.2. The relative orientations of successive molecules in the alternating sequence is neither fully staggered nor eclipsed. There are alternating planes composed of only benzene and hexafluorobenzene molecules; with the plane of the benzene rings tilted away from the c-axis

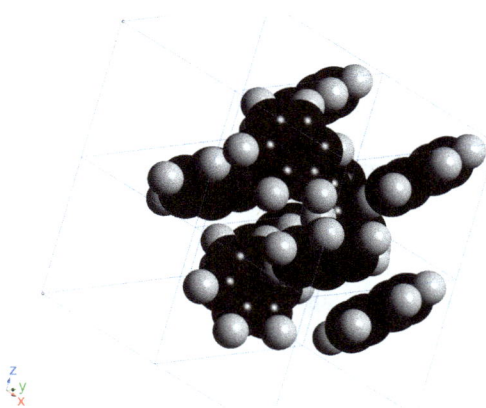

**Figure 11.1.** The structure of solid benzene, as determined by x-ray scattering; carbon atoms are black and the hydrogen atoms are white. The molecules are represented as space-filling; that is, the structure is calculated from the experimental data using appropriate vales for the sizes of the carbon and hydrogen atoms, and we see how these molecules fill space.

**Figure 11.2.** The structure of the lowest temperature phase (IV) of benzene:hexafluorobenzene; this phase exists below 205 K (carbon atoms are black, hydrogen atoms are white and the fluorine atoms are green).

by about 23° and by 27° for the hexafluorobenzene rings. This structure belongs to the monoclinic system and is characterized by the following unit cell lengths: $a = 9.4952$ Å, $b = 7.4236$ Å and $c = 7.5260$ Å ($\beta = 95.63°$); the occupancy ($Z$) of the unit cell is 4.

The adduct, benzene:hexafluorobenzene, is the simplest member of a large class of layered organic compounds. The adduct melts at a higher temperature than the solids composed of the pure components. However, it is a structure which displays structural instabilities or phase transitions [3], where the arrangement of the molecules in the solid changes at well-defined temperatures. Whereas pure solid benzene and pure solid hexafluorobenzene do not undergo any structural phase transitions below their melting points, in the adduct there are three phase transitions at −68 °C, −26 °C and 2 °C. That is, the solid state structure of the adduct is unstable to fluctuations in thermal energy. The question then becomes, what drives these structural changes that are seen in the binary complex but not in the solids formed from either benzene or hexafluorobenzene?

In figures 2.4 and 11.2 we see the structure of solid benzene:hexafluorobenzene in its lowest temperature phase (as determined by a combination of x-ray and neutron scattering). We see that this binary solid may be described as being composed of alternating layers of benzene and hexafluorobenzene molecules, this being the most stable configuration for the electrostatic interaction of electric quadrupole moments of opposite sign. Thus, the structure of the binary solid is a series of parallel stacks of alternating benzene and hexafluorobenzene molecules.

What we also observe in figures 2.4 and 11.2, is the interpenetration of the C–H and C–F bonds of molecules in neighbouring columns. Thus, at temperatures below −68 °C the bonds on one type of molecule closely approach neighbouring molecules of the other type. Thus C–H bonds on a benzene molecule are directed towards, and come very close to F atoms of the C–F bonds of a hexafluorobenzene molecule in the

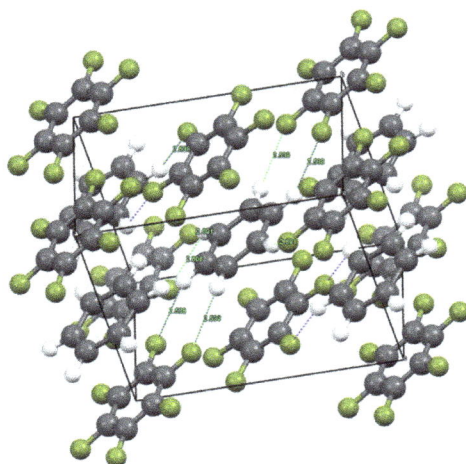

**Figure 11.3.** Internal structure and close approach between the molecules in the lowest temperature phase (IV) of benzene:hexafluorobenzene; the fluorine atoms are green, the carbon atoms are grey and the hydrogen atoms are white.

layers above and below. It is possible, therefore, to envisage a network of weakly polarized hydrogen-like bonds as the means of stabilizing the solid lattice. Given that a hydrogen bond is a well described entity that involves the sharing of a hydrogen atom between two electronegative atoms, it is perhaps better to describe these $(\delta-)$C–H$(\delta+)\cdots(\delta-)$F–C$(\delta+)$ interactions as bond dipole–bond dipole interactions. If we say that $\mu_{C\text{-}H}$ and $\mu_{C\text{-}F}$ are the two bond dipole moments, then the intercolumn interactions in this solid will vary as $\mu_{C\text{-}H}\,\mu_{C\text{-}F}/r^3$; where $r$ is the separation of the bond dipole moments.

This complex interpenetration of the C–H and C–F bonds in benzene:hexafluorobenzene is well seen in figure 11.3, where we see how the C–H$\cdots$F–C interactions hold together the layers and the closely-packed columns that constitute the solid. We are looking at complex, crystal specific interactions. The dashed blue lines correspond to the closest C–H$\cdots$F–C distances; that is, to the shortest and hence the strongest of the bond dipole moment–bond dipole moment interactions that hold the close packed arrangement of columns together, and which therefore account for the observed complex dynamics seen in this solid. In the image, there are two closest distances to neighbouring F-atoms for the six hydrogen atoms in a particular benzene molecule; four hydrogen atoms are separated from their nearest F-atom by 2.538 Å and the other two H-atoms are separated by 2.691 Å from their nearest F-atoms. This difference will have a significant effect on interactions perpendicular to the column axis as the dipole–dipole interaction scales as $r^{-3}$ (where $r$ is the separation). The hydrogen bond-like bond dipole–bond dipole interactions are clearly seen to hold the layers of the solid together, by binding the closely-packed columns of molecules.

Molecules such as benzene and hexafluorobenzene have no permanent electric dipole moment. However, they do have bond dipole moments. There are six of them oriented symmetrically, and the vector sum of their values is zero[1]. It is through the intermeshing network of C–H and C–F bonds (C–H···F–C interactions) originating in the different columns that the columns of molecules are stabilized, the layers are weakly bonded to each other, and thermal energy is transmitted through the crystal. This low temperature phase of the solid adduct is very much like a molecular *gearbox* with interlocking wheels (the interpenetrating C–H and C–F bonds) on parallel shafts (or columns of molecules) allowing the transfer of energy throughout the crystal. The thermally activated motion of these bonds passes from one column to neighbouring columns because of the closeness of approach of the bonds. They are thermally driven, oscillating bond dipole moments, and the electromagnetic fields that they emit when modulated by thermal excitations are sensed by their neighbours. Research has shown [4] that the phase transition at −68 °C is like a molecular clutch being engaged in a transmission system to change a gear wheel; in the solid, the molecular clutch leads to an increase in the intercolumn spacing. The columns have separated, thus facilitating further rotations of the benzene and hexafluorobenzene molecules, which continues right up to the melting point at 25 °C. Closely related molecules such as toluene and ethylbenzene cannot follow this relaxation route and so they melt.

Whereas in solid benzene, electric quadrupole moments of like-polarity interact to give a solid composed of slipped parallel and *T* ordered pairs of molecules, in benzene:hexafluorobenzene, the opposite polarity of the quadrupole–quadrupole interaction produces stacked columns of alternating, parallel benzene and hexafluorobenzene molecules. At low temperatures, the columnar structure gains an additional stability from the transverse, or intercolumnar bond dipole moment–bond dipole moment interactions. These stabilizing intermolecular interactions, scaling as $r^{-3}$ (where $r$ is the inter-dipole separation, see figure 11.3), are equivalent to the bond dipole–quadrupole intermolecular interactions, scaling as $r^{-4}$, that stabilize solid benzene; see equation (2.5). However, this structure is less stable to thermal excitations than the arrangement in solid benzene; the one structure being more stable to thermal excitations, particularly, excitations transverse to the main direction of the columns of molecules, than the other.

## 11.2 Thermal expansion

Figure 11.4(a) displays a simple model for the structure of the adduct benzene: hexafluorobenzene; that is, parallel columns comprising alternating, equi-spaced benzene and hexafluorobenzene molecules. This material has been investigated by both neutron and x-ray diffraction, and in figure 11.4(b) we display neutron scattering data collected on benzene:hexafluorobenzene over a range of temperatures, where we see clearly the phase transition at 205 K. Indeed, we see that the two phases co-exist.

---

[1] In the same way that we may envisage a linear quadrupole moment (such as $CO_2$) to be a series of partial charges, $(\delta-)O-C(2\delta+)-O(\delta-)$, we may imagine a benzene rings as having $6\delta+$ charges distributed over the six C–H bonds and the balancing $6\delta-$ partial charges above and below the plane of the molecule (see red lettering in figure 10.1). This is a distributed multipole model of the interacting charge distributions.

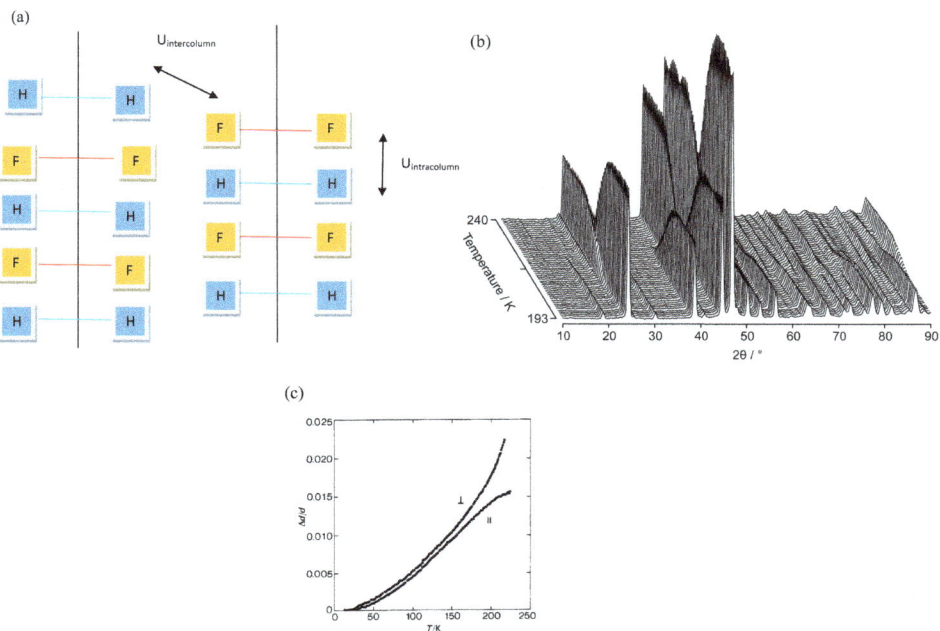

**Figure 11.4.** (a) Schematic representation of the dynamics of adducts such as benzene:hexafluorobenzene, or mesitylene:hexafluorobenzene (hydrogenated molecules in blue, and fluorinated molecules in yellow). We are looking perpendicular to the $c$-axis of the crystal, and for simplicity we have shown the hydrogen-containing aromatic molecules (molecular quadrupole moment < 0) and the hexafluorobenzene (molecular quadrupole moment > 0) to be ordered perpendicular to the $c$-axis. There are two electrostatic interactions of interest; $U_{intercolumn}$ is the bond dipole–bond dipole interaction between C–H and C–F bonds, C–H and C–H bonds, and C–F and C–F bonds in molecules situated in adjacent columns, and $U_{intracolumn}$ which is the quadrupolar electrostatic interaction between parallel, nearest neighbour molecules in the same column. (b) Diffraction data for neutrons of wavelength 2.52 Å scattered by $C_6D_6$:$C_6F_6$ as a function of temperature between 193 and 240 K, clearly demonstrating the phase transition at 205 K. (c) Linear thermal expansion $\Delta d/d$ of $C_6D_6$:$C_6F_6$ as a function of sample temperature; for intracolumn (global) ($\perp$) and intracolumn (global) ($\parallel$) spacings.

The detailed structure of benzene:hexafluorobenzene is given above, here we point out that the crystal is composed of closely-packed columns of alternating and equi-spaced benzene and hexafluorobenzene molecules, which are parallel to the $c$-axis of the crystal; the intermolecular spacing changes with temperature. The intermolecular spacing parallel and perpendicular to the column axis were derived from the scattering data shown in figure 11.4(b); from the (001) and (110) reflections, respectively. In figure 11.4(b) and figure 3.2 (Phase IV), the small peak at 19° is the (001) reflection, the large peak at 24.5° is the (110) reflection, the peak at 39.5° is the (002) reflection and the biggest line at 49° is the (−202) reflection.

The linear thermal expansion data derived from the diffraction data in figure 11.4(b) is given in figure 11.4(c), and we will now see how useful such information can be in understanding the architecture of such crystals. Although thermal expansion depends upon the intermolecular forces that maintain the structure of the solid under investigation, it is not generally possible to invert measured expansion coefficients to yield useful values for the intermolecular potentials. However, the solid being examined here is a closed-packed set of columns with equi-spaced molecules running along each column, as given in the model in figure 11.4(a). Consequently, it is possible to model an intercolumn and an intracolumn potential. This is possible in this material, because of the strongly anisotropic nature of the structure; closely-packed columns. We will see that measurements of the thermal expansion in the directions parallel and perpendicular to the column axis can provide useful information to those seeking to design crystals; in much the same way that we have seen how high-resolution laser Raman spectroscopy also provides information about the evolution of the intermolecular potentials within the solid as the temperature varies.

We assume that the thermal expansion is proportional to the internal energy, which is equivalent to assuming that each vibrational mode (of frequency $\omega$) makes a contribution proportional to its frequency, and is consistent with the relation due to the German physicist Eduard Grüneisen (1877–1949):

$$\Delta d = (\mathrm{c}h/2\pi k_B) < \omega[\exp(h\omega/2\pi k_B T) - 1]^{-1}> \tag{11.1}$$

Here, c is a constant, defined below, h and $k_B$ are the constants due to Planck and Boltzmann, respectively. The linear thermal expansion, $\Delta d$, represents the difference in lattice spacing per molecular unit over the temperature range. The angle brackets denote a thermal average over the vibrational states, which contribute to the expansion in a particular direction. Unfortunately, it is this averaging that usually nullifies any attempt to use this theory to explore intermolecular potentials in isotropic crystals.

However, in benzene:hexafluorobenzene at high-temperature we may expand equation (11.1) to give

$$\Delta d \approx c\{T - (h/2\pi k_B) < \omega > /2\}. \tag{11.2}$$

In order to estimate the constant, $c$, we approximate the intermolecular potential. For example, we may approximate the crystal in the high-temperature, classical regimen to be an ensemble of anharmonic Morse oscillators, vibrating with a fundamental frequency $<\omega>$. The properties of the Morse oscillator are given in [4], and $c$ is given by

$$c = 3k_B/4A\alpha \tag{11.3}$$

where $A$ and $\alpha$ are the Morse well-depth and asymmetry parameter, respectively. Specializing to a Lennard-Jones 6-12 potential, gives $\alpha \approx 6/d$ where $d$ the spacing between the benzene and hexafluorobenzene molecules running along a column. Combining these two approximations gives

$$\Delta d/d = (1/8U)(T - \Omega/2), \tag{11.4}$$

where $\Omega = h<\omega>/2\pi k_B$ is the average frequency and $U = A/k_B$ is the strength of the bonding between the molecules (both expressed as temperatures).

In the solid being considered here, $d$ is the centre-of-mass to centre-of-mass distance between the parallel benzene and hexafluorobenzene molecules.

The simple expression (11.4) describes the data for $C_6D_6{:}C_6F_6$ rather well, over the temperature range 80–150 K, as seen in figure 11.4(c). There is no appreciable anisotropy between $\Delta d/d$ measured perpendicular and parallel to the direction of the molecular chain, the $c$-axis. In both cases, $U \approx 1300$ K (or 904 cm$^{-1}$) and $\Omega \approx 100$ K (or 69.6 cm$^{-1}$). The density of states for this material, measured by inelastic neutron scattering is given in the upper part of figure 10.3(a). And there is good agreement between our approximated values of the density of states and the actual density of states.

From our modelling, the Morse/Lennard-Jones fundamental frequency is given by:

$$\omega_0 = (6/d)\sqrt{(2A/m)} \tag{11.5}$$

where $m$ is the reduced mass of a benzene-hexafluorbenzene pair. Inserting values of these quantities given above, gives $h\omega_0/k_B \approx 75$ K (or 52 cm$^{-1}$), which agrees well with the lowest frequency intermolecular vibrations observed in this material (see the upper part of figure 10.3(a), which displays the density of states of this material in the frequency range 0–160 cm$^{-1}$).

Such investigations of the expansion of organic materials, with a view to determining the dynamics and interactions of the molecules within the solid are not often possible because of the complexity of the crystal architecture. The adduct benzene: hexafluorobenzene is composed of closely-packed chains of parallel benzene and hexafluorobenzene, but solid benzene contains the classic herringbone type pattern of molecular packing where there are two orientations of the planar molecules; slipped parallel and T-shaped. Consequently, the type of analysis given above is more complex than for benzene:hexafluorobenzene, and so more approximations have to be made to achieve any form of analysis, and so less useful information about intermolecular forces and the role they play in crystal engineering is forthcoming.

This simple, yet informative model about the dynamics of molecules in crystal architectures is consistent with the results of quasielastic neutron scattering measurements [3]. Above 150 K, benzene:hexafluorobenzene is observed to quasielectically scatter neutrons, and this scattering may be interpreted as the onset of a thermally-activated 6-fold rotation of the benzene rings, about their molecular axis, within the solid. From the neutron scattering experiments, the barrier to activation of this rotation was estimated to be about 9 kJmole$^{-1}$ $\approx 1100$ K. This experimental figure is the same order of magnitude as the intermolecular bond strength, 1300 K, derived from the modelling of the thermal expansion of the crystal, demonstrating the utility of the modelling. Taking the known vales of the molecular quadrupole moments of these molecules and the known ring separation of about 3.6–3.7 Å, we calculate that the attractive potential, along a chain, between a benzene and a hexafluorobenzene molecule, which would have to be overcome for a rotation of one of the molecules to take place is equivalent to the amount of energy represented by the temperature 854 K.

Thus, information from thermal expansion experiments and quasielastic neutron scattering give us an insight into crystal engineering. These experiments tell us that at temperatures about 150 K, in crystals of adducts like benzene:hexafluorobenzene, there is an onset of a hindered rotational motion of the benzene molecules around their $C_6$ rotational axis. And this rotational motion provokes a reorganization of the crystal structure. It is likely, given the temperature dependence of this hindered motion that both the benzene and the hexafluorobenzene molecules are rotating freely about the $c$-axis of the crystal at higher temperatures. And that the three phase transitions seen in this material below its melting temperature (298 K) are the onset of such free rotation. A free rotation which is the origin of the considerable diffuse scattering observed in the neutron diffraction pattern of this material (see figure 8.5).

## 11.3 Mesitylene:hexafluorobenzene

Understanding weak non-covalent interactions is crucial for the prediction and control of organic solid-state structures. Of particular interest is the design of organic co-crystals as alternatives to salts in the development of new materials in, for example, the pharmaceutical industry. For organic fluorine (i.e. bonded as C–F), there is a general consensus that fluorine rarely forms hydrogen bonds, leading to questions about the nature of the interaction between C–F and H–C (or $H_3C$–) in the solid state.

One of the simplest organic co-crystals containing a molecule with a C–F bond and without hydrogen bonding is the adduct of benzene and hexafluorobenzene, which we looked at earlier. Here we will consider the structure of the adduct formed between 1,3,5-trimethylbenzene (mesitylene) and hexafluorobenzene.

The quadrupole moment of mesitylene is larger than that of benzene at $-32.0 \times 10^{-40}$ C m$^2$, due to additional positive electric-charge residing on the $-CH_3$ groups. Thus, it is not surprising that mesitylene also forms a binary adduct with hexafluorobenzene due to the favourable interaction of quadrupole moments as in benzene:hexafluorobenzene. In benzene:hexafluorobenzene, rotation of the rings occurs about the high symmetry axes of each component and begins at about 150 K. By contrast, the methyl groups of mesitylene can be expected to hinder such rotation. However, the methyl groups of mesitylene are free to rotate in order to optimize more specific and local charge interactions.

The structure of mesityle:hexafluorobenzene has recently been solved by powder x-ray diffraction [5]. The data revealed the presence of two phase transitions; at 111 K and at 179.2 K. The structures of phases I, II, and III have been solved by a combination of powder and single crystal measurements, and a summary of the space-group symmetry and unit cells for each phase may be found in [5]. Even though there are two phase transitions below the melting point of this solid, there is no large anisotropic expansion and the volume of the unit cell is seen to increase almost linearly from 347 cubic angstroms at 100 K (phase III) to 376 cubic angstroms at 300 K (phase I). The structure of these three phases of this material are given in figure 11.5.

In all three phases, the molecules are arranged in columns formed of alternating hexafluorobenzene and mesitylene molecules arranged face to face. The columns themselves are approximately close-packed with six columns surrounding each

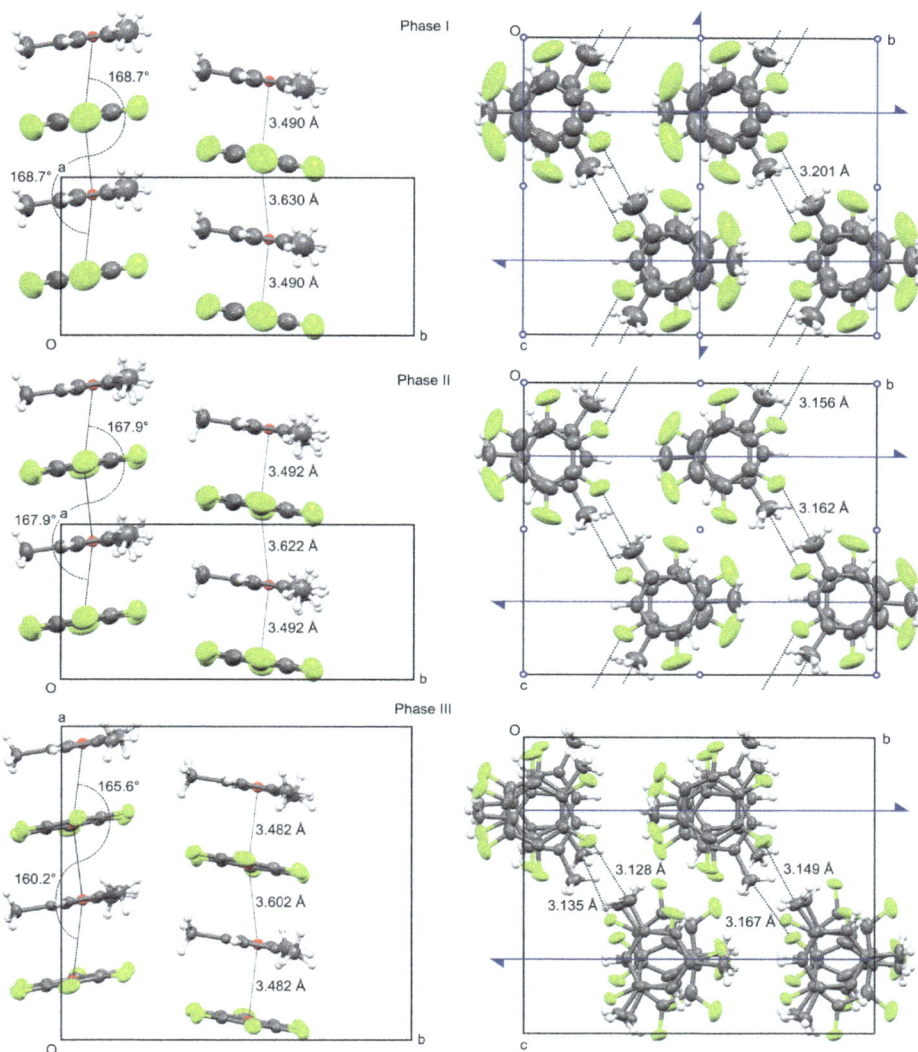

**Figure 11.5.** Left: view of the crystal structures of mesitylene:hexafluorobenzene seen down the *c*-axis for phase I (top), phase II (middle), and phase III (bottom) showing the distances between ring centroids shown in red and the angles formed by lines joining the ring centroids. In phase III, there are two crystallography distinct pairs of molecules: an additional set of inter-planar distances (3.482(3) Å and 3.648(3) Å) and corresponding angles (155.6(1)° and 168.3(1)°) were calculated from the second pair (not seen in bottom-left figure). The increase in the dynamical motion of the molecules with increasing temperature is evidenced by the size of the *thermal* ellipsoids. C, F, and H atoms types are drawn in grey, green, and white, respectively. Right: views seen down the a-axis for phase I (top), phase II (middle), and phase III (bottom) showing: (i) the staggered arrangement of the rings within a column; (ii) the C–F⋯CH₃–C interactions that bind adjacent columns together; and (iii) selective symmetry elements present in each crystal structure (blue).

column. It is noteworthy that in all three phases: (i) the molecules are slightly tilted with respect to the column axis (aligned to the crystallographic $a$-axis); (ii) they exist as distinct pairs with alternating ring-to-ring centroid distances of about 3.5 Å and 3.6 Å; (iii) within a column, the hexafluorobenzene and mesitylene rings are staggered, similar to the behaviour of the rings observed in phase IV of benzene: hexafluorobenzene, and (iv) that the binding between adjacent columns occurs via weak electrostatic interactions between the C–F bond dipole moment in one column and the $H_3C$–C bond dipole moment in neighbouring columns.

The variation in the influence of this intercolumnar interaction on the crystal structure, as a function of temperature gives rise to the three solid phases. Positions of all hydrogen atoms in mesitylene can be located by single crystal diffraction studies in all three phases of mesitylene:hexafluorobenzene. In phase I, all three – $CH_3$ groups of mesitylene appear crystallographically disordered; in phase II, two of the –$CH_3$ groups are disordered but one is ordered; and in phase III, all three of the – $CH_3$ groups are ordered.

The measured near linear arrangement of the positions of the heavy atoms in $F \cdots H_3C$–C suggests an intercolumnar interaction that has little dependence on the orientation of the H atoms of the methyl group about the $H_3C$–C bond. This electrostatic interaction is sufficiently strong to pin down two of the F atoms of an adjacent hexafluorobenzene ring and explains the observed thermal motion of the hexafluorobenzene ring as evidenced by the large variation in anisotropic displacement ellipsoids for the F atoms in all three phases (figure 11.5). The third methyl group in mesitylene is unable to align directly with any F atom and in phase II its H atoms adopt a fixed position resulting in a slight displacement of adjacent columns parallel to the column axis, and a monoclinic distortion of the structure. Further ordering of the methyl groups occurs on cooling to phase III as further optimization of the intercolumn interactions takes place. In all three phases, the net in-plane librational motion of the rigid hexafluorobenzene molecule is not about its centre of mass in contrast to the averaged librational cloud of the mesitylene molecule, where the rotational axis is about its centre.

Again, we see how simple ideas of molecular electrostatics, based on the observed values of the molecular quadrupole moments, allow us to explain the ordering of molecules within the close-packed stacked columns present in the structures of all three phases; and the concept of a bond dipole moment permits an explanation of the shorter-range intercolumnar interactions. These interactions become increasingly constrained as the temperature is decreased and are the origin of the phase-transitions.

## 11.4 To dimerize, or not to dimerize...

The molecules interacting in benzene:hexafluorobenzene and mesitylene:hexafluorobenzene have no permanent electric dipole moments, but they are subject to intermolecular forces arising from their electrical properties. Modelling the structure of the columns of alternating benzene and hexafluorobenzene molecules interacting via the attractive force generated by their quadrupole moments allows one to begin to unravel the rich dynamics of these solids composed of close-packed columns. But what of the intercolumnar interactions, which play a predominant role in the origin of the observed phase transitions? Indeed, the structure of these solids is the result of the balance of intracolumnar interactions and the interaction between the close-packed columns; the intercolumnar interactions, see figure 11.4(a). And it is likely that whether or not distinct dimerization, observable via x-ray diffraction, occurs in these adducts rather than a vibrational dynamic dimerization, or Peierls's distortion occurs along the $c$-axis will depend upon the relative strength of the intercolumnar and intracolumnar interactions.

We now come to the central issue of this investigation of the structure of the layered binary adducts benzene:hexafluorobenzene and mesitylene:hexafluorobenzene, which are only two representatives of a very large class of compounds. Why in benzene:hexafluorobenzene are the individual benzene and hexaflorobenzene distributed equidistantly along the column axis that defines the solid; that is; there is no evidence of dimerisation seen by x-ray diffraction studies, but in mesitylene:hexafluorobenzene the two molecules that constitute the solid are seen to form distinct dimers and it is these dimers that are distributed along the column axis. Referring back to figure 2.1, in benzene:hexafluorobenzene, $R = R_{12}$, but in mesitylene:hexafluorobenzene, $R$ and $R_{12}$ are measurably different; why is this? Both solids are held together by intermolecular forces of the same magnitude, the quadrupole moment of mesitylene is about 6% larger than that of benzene, both solids melt at about the same temperature (benzene:hexafluorobenzene melts at 25 °C and mesitylene:hexafluorobenzene melts at 30 °C), and both solids display structural instabilities, or solid-state phase transitions.

It is likely that the explanation of why some adducts are dimerised and some are not, is to be found in the relative magnitudes of the intercolumnar (bond dipole–bond dipole) interactions and the intracolumnar (molecular quadrupole–molecular quadrupole) interactions. These hydrogen-bond-like intercolumnar links are clearly seen in figure 11.3 for benzene:hexafluorobenzene. In this adduct, any one benzene ring is linked, or held in place in the lattice, by four close bond dipole moment–bond dipole moment interactions. It is seen that in that benzene molecule, the 2, 3 and 5, 6 positions are linked to hexafluorobenzene molecules in the molecular layers above and below it by particularly short C–H...F–C bonds. This leaves the 1 and 4 positions on the benzene ring attached by weaker bonds to neighbouring hexa-fluorobenzene molecules. It is the number of these short intercolumn links that will be key to deciding whether or not the adduct displays distinct dimerization.

## 11.5 *S*-Triazine

Symmetric-triazine is deliquescent white solid that melts at 81 °C. It was first prepared in 1895, but it was not until the middle of the last century that the detailed structure of the molecule was determined. Indeed, it was only in 1937 that Linus Pauling demonstrated theoretically that the molecule could exist at all. The molecule is shaped like benzene, .

The room temperature crystal structure is given in figure 11.6. The unit cell is hexagonal with cell dimensions $a = b = 9.647$ Å and $c = 7.281$ Å. The 3-fold axes of

(a)

(b)

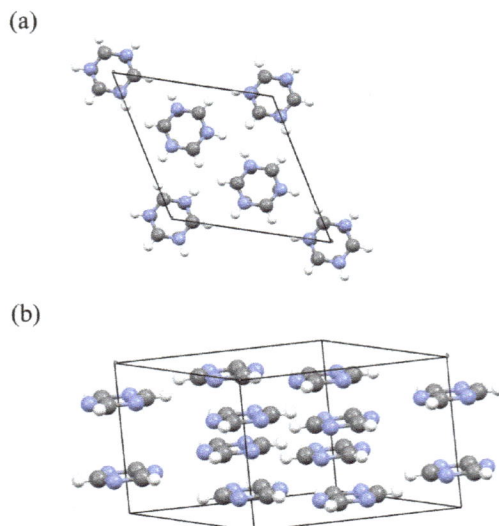

**Figure 11.6.** Two views of the crystal structure of the high-temperature phase of *s*-triazine; the carbon atoms are grey, the hydrogen atoms are white and the nitrogen atoms are blue. In (a) we are looking down the *c*-axis at columns of molecules. In (b) we see the equi-spaced pairing of the molecules along the columns; each pair of *s*-triazine molecules represents a staggered arrangement of nitrogen atoms; the N-atoms in one molecule are paired with the C-atoms in its partner.

the molecules occupy crystal sites of similar symmetry with the result that infinite stacks of parallel molecules are formed, the molecular planes being perpendicular to the *c*-axis. Adjacent molecules within these stacks are related by a centre of symmetry which is equivalent to a relative rotation of 60° about the symmetry axis combined with a translation of *c*/2. Notice the absence of any herringbone-like orientation of the molecules that would have indicated that this structure is determined by quadrupole–quadrupole interactions as in benzene. This structure is more reminiscent of the structure of benzene:hexafluorobenzene (see figure 2.4) than benzene (figure 2.3); even though it is a pure solid not an adduct.

The molecule is planar and aromatic, and so should have similar properties to benzene; for example, given its aromatic character, *s*-triazine should have a negative electric quadrupole moment with the charge clouds from the three carbon atoms distributed above and below the plane of the hexagonal molecule. The quadrupole moment of *s*-triazine has been estimated to be about $3 \times 10^{-40}$ C m$^2$. Given that the molecule is aromatic, the near zero value of its quadrupole moment implies that the contributions of the alternating C–H bonds and N atoms in the planar ring cancel each other's contributions to the overall electron distribution, and we are left with a molecular quadrupole moment, of order, that of 1,3,5-trifluorobenzene, where the same cancelling effect (a demonstration of the bond-additivity of molecular properties) has been observed.

However, whereas benzene is a regular hexagon, *s*-triazine displays a marked departure from regularity. Neutron scattering studies revealed that the C–N and

C–H bonds lengths are 1.315 Å and 0.93 Å, respectively, and the N–C–B bond angle is 125°. The hexagonal unit cell of $s$-triazine has $Z = 6$, and the molecule occupies Wyckoff position 2(a) of site symmetry 32; the space group is R3c. Close inspection of this structure, as seen in figure 11.6(b), reveals that the interactions between the columns of $s$-triazine molecules are a network of hydrogen bond-like interactions between C–H bonds on one molecule and the lone pairs of the N atoms in molecules in neighbouring columns. Each of the three C–H bond dipoles is closely directed at a N atom in a neighbouring molecule. Thus, the planes of $s$-triazine molecules are stabilized by an intracolumnar network of weak hydrogen bonds. In the case of $s$-triazine, because the quadrupole moment is small it is the intercolumnar hydrogen bonds that are particularly important in maintaining the high temperature phase.

Consequently, the structure of this crystal at ambient temperature is like that of the binary adduct benzene:hexafluorobenzene, but dominated by intercolumnar rather than intracolumnar interactions. The structure of the higher temperature phase of $s$-triazine is not at all like the structure of benzene, which has the classic herringbone arrangements of quadrupole moments of the same phase.

Crystalline $s$-triazine undergoes a phase transition to a monoclinic structure (space group C2/c) at 198 K. The crystal structure of $s$-triazine at 150K has been studied by x-ray diffraction; the unit cell has dimensions: $a = 6.884$ Å, $b = 9.569$ Å, $c = 7.093$ Å, beta $=126.61°$ (very like that of solid benzene, which is orthorhombic with: $a = 7.46$ Å, $b = 7.034$ Å and $c = 9.666$ Å, and Z = 4), but the structure of this low temperature phase of $s$-triazine has not been solved, so the detailed locations of all of the atoms in this lower temperature phase and their relative orientations are unknown.

Given the information that we do possess about this lower temperature phase, it is not composed of columns of molecules and may well have the slipped parallel arrangements of $s$-triazine molecules as seen in solid benzene, in which case the crystal architecture will be stabilized by bond dipoles (C–H on one molecule) interacting with the N atoms and $\pi$-cloud on neighbouring molecules.

## 11.6 Naphthalene

Naphthalene is a naturally occurring hydrocarbon derived from fossil fuels, specifically coal tar, which may contain up to 10% naphthalene. It is a white solid with a strong, characteristic odour that sublimes readily at 80 °C. Naphthalene was among the first organic molecules to be studied with x-ray crystallography; with W H Bragg undertaking the first studies in the early 1920s. The structure of crystalline naphthalene was first determined in 1928.

Naphthalene has the chemical formula $C_{10}H_8$, and consists of two benzene molecules joined along one edge. This makes it the simplest of the polyaromatic hydrocarbons, a class of molecules made up of multiple fused aromatic carbon rings. And consequently, many of the properties of naphthalene may be predicted from a consideration of the properties of benzene; for example, the electric quadrupole moment of benzene is $-29 \times 10^{-40}$ C m$^2$, and that of naphthalene is $-46 \times 10^{-40}$ C m$^2$,

(a)

(b)

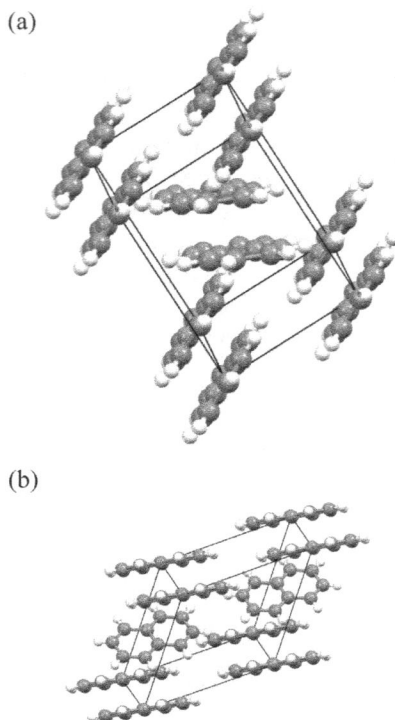

**Figure 11.7.** Two views of the crystal structure of naphthalene, $C_{10}H_8$; the carbon atoms are grey and the hydrogen atoms are white. Part (a) clearly demonstrates the classic herringbone structure that is displayed by structures formed primarily by the interaction of molecular quadrupole moments.

and that of anthracene, which is three benzene rings joined in a line is $-61 \times 10^{-40}$ $Cm^2$ (see section 9.1). Similarly, the polarizability of these molecules scale in proportion to the size of the molecule[2].

These similarities in properties will also be reflected in the type and magnitude of the attractive and repulsive forces that operate between these molecules in the condensed phase, and that generate the observed crystal architectures for these materials. Figure 11.7 give two representations of the structure of naphthalene. These figures may be compared with figures 2.3 and 11.1, which give the structure of benzene. The similarity in the packing between these two materials is clear. And given that the structure of solid benzene may be rationalized in terms of the attractive part of the interaction of two large quadrupole moments of the same phase, we may assume that a similar analysis would explain the structure of solid naphthalene. However, it needs to be remembered that it is also necessary to consider the interactions between the C–H bond dipoles on one benzene molecule

---

[2] In the c.g.s. system of units, the molecular polarizability has units of volume, of order, $10^{-24}$ $cm^3$. In the SI, the units of polarizability are $C^2$ $m^2$ $J^{-1}$; the conversion factor is $4\pi\varepsilon_0$.

and the $\pi$-cloud on adjacent molecules to fully explain the observed structure, and these interactions are essential to explain the internal dynamics of the solid. Similarly, with naphthalene we need to consider the edge–face interactions between the molecules and how this contributes to the observed crystal structure.

One must also be aware that a naphthalene molecule is considerably more asymmetric that a benzene molecule, and this asymmetry will certainly contribute to the engineering of the crystal. This difference is seen in the lattice parameters of the two solids. The crystals of naphthalene belong to the monoclinic system of crystals and have the following lattice parameters: $a = 8.2606$ Å, $b = 5.987$ Å, and $c = 8.6816$ Å with $\alpha = 90°$, $\beta = 122.92°$, and $\gamma = 90°$; at 290 K. The crystal structure of benzene is orthorhombic, of dimensions $a = 7.46$ Å, $b = 7.034$ Å and $c = 9.666$ Å, with $Z = 4$; see table 11.1 for more details.

The contribution of the length of polyaromatic hydrocarbons, which can be considered as a repulsive, short-range interaction on the packing of these molecules together to form solids may be estimated by studying the structure of anthracene, tetracene and pentacene. For example, in figure 11.7(a), we see the structure of naphthalene, which demonstrates the classic herringbone arrangement of two inequivalent molecules. We immediately see, on making a comparison with the structure of benzene, that the T-shaped type of interaction seen in solid benzene (as seen in figures 2.3 and 11.1) has become modified due to the length of the longer molecule, to become less of a perpendicular edge–face interaction.

In figure 11.8, we see the evolution of the herringbone structure displayed by interacting aromatic (quadrupolar) molecules for increasing size of molecule. In fact, the image displays the structure of benzene, which we have seen before, and pentacene, that is, five benzene molecules fused edge to edge. What is seen is that as the molecules get longer, that is, as short-range interactions become increasingly important, the T-shaped orientation of the interacting quadrupole moments (clearly seen in figures 2.3 and 11.1 for benzene) are modified. The angle between the planes

**Table 11.1.** Details of crystal structures of increasingly large polynuclear aromatic molecules.

| Molecule | Crystal system and occupancy | $a$/Å | $b$/Å | $c$/Å | Angle in herring-bone structure/ degrees |
|---|---|---|---|---|---|
| Benzene $C_6H_6$ | Orthorhombic $Z = 4$ | 7.29 | 6.74 | 9.47 | 85 |
| Naphthalene $C_{10}H_8$ | Monoclinic $Z = 2$ | 8.29 | 5.97 | 8.68 | 53 |
| Anthracene $C_{14}H_{10}$ | Monoclinic $Z = 2$ | 8.561 | 6.036 | 11.163 | 51 |
| Tetracene $C_{18}H_{12}$ | Triclinic $Z = 2$ | 7.98 | 6.14 | 13.57 | 53 |
| Pentacene $C_{22}H_{14}$ | Triclinic $Z = 2$ | 7.93 | 6.14 | 16.03 | 53 |
| Hexacene $C_{26}H_{16}$ | Triclinic $Z = 2$ | 7.9 | 6.1 | 18.4 | 52 |
| Coronene $C_{24}H_{12}$ | Monoclinic $Z = 2$ | 16.11 | 4.7 | 10.10 | 44 |
| Graphite | Hexagonal $Z = 4$ | 2.461 | 2.461 | 6.708 | 0 |

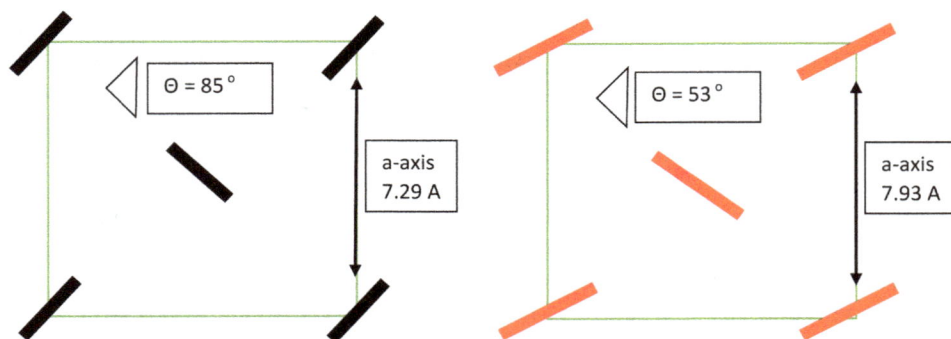

**Figure 11.8.** The evolution of the herringbone orientation of aromatic molecules in the solid state. The left-hand image is the structure of the unit cell of benzene looking down the *c*-axis; the planar molecules are seen *side on*. The right-hand image is the same view but of pentacene. Of particular importance is the change in the angle of what in benzene is the T-shaped orientation and which is clearly seen in figure 2.3.

of the two interacting molecules falls. Thus, the planes of the two benzene molecules (see figure 2.3) are almost perpendicuar at 85.2°, but in pentacene this angle has fallen to 52°, and in coronene (six benzene rings arranged in a hexagon, $C_{24}H_{12}$) the angle has become 44°. Table 11.1 contains the details of the changing unit cell dimensions for all these polynuclear aromatics. Indeed, in the limit of polynuclear aromatic solids, that is, graphite, this angle has become zero and there are only parallel layers of interacting carbon atoms; we have lost the T-shaped orientation of interacting quadrupole moments.

From table 11.1 we see how the unit cell parameters vary on going from benzene to hexacene and coronene. It should be noted, however, that direct comparison of the distances given is not appropriate as these structures are in different crystal systems. But we do see that as the molecules become larger and have an intrinsic axis within the molecule, we move from an architecture that displays a classic herringbone form in benzene to one that is distorted, and in the limit of *molecule* size, to a structure of parallel layers, as in graphite[3]. The central perpendicular molecule in the benzene structure having moved to slot in between the parallel molecules that are on the corners of the unit cell. And one might suggest that as the central molecules move into position between the two molecules at the unit cell corners, the resulting parallel layers are closer together. In this image, the *a*-axis distance has halved, and we are now closer to the separation of 3.35 Å of the parallel layers of carbon seen in graphite. We have moved from discrete molecular interactions in covalent crystals to a giant covalent structure.

# Further reading

[1] Cox E G, Cruickshank D W J and Smith J A S 1958 *Proc. R. Soc.* A **247** 1
[2] Boden N, Davis P P, Stem C and Hand Wesselink G A 1973 *Mol. Phys.* **25** 81–6

---

[3] If anything, it is benzene that is the odd one out in this series... in the same way that the elements of the first row of the periodic table have properties that are not shared by the other members of their groups.

[3] Williams J H 1993 *Chem. Phys.* **172** 171–86; Williams J H 1991 *Mol. Phys.* **73** 99–112; Williams J H 1991 *Mol. Phys.* **73** 113–25

[4] Bramwell S T and Williams J H 1992 *J. Chem. Soc. Faraday Trans.* **88** 2721–4

[5] Cockcroft J H, Ghosh R E, Shephard J J, Singha A and Williams J H 2017 *Cryst. Eng. Comm.* **19** 1019

# Chapter 12

## Supra-molecular chemistry

The previous chapters have dealt with solids formed from small molecules. The solids formed from such small polyatomic molecules reflect this, and the properties of the solid arise from the properties of the constituent molecules. We will now look briefly at supra-molecular architectures, where the molecules that make up the solids may be relatively small, but they are first built into larger units and it is from these larger units that the final crystal is constructed. This is probably one of the most active fields of modern synthetic chemistry.

### 12.1 Metal–organic frameworks

Metal–organic frameworks (MOFs) are compounds consisting of metal ions, or clusters of metal ions coordinated to organic ligands to form 1-, 2-, or 3-dimensional structures. They are a subclass of coordination polymers, with the special feature that they are often porous. The organic ligands used in the synthesis of MOFs are sometimes referred to as struts; examples being 1,4-benzenedicarboxylic acid or oxalic acid. These are ligands capable of binding two metal atoms, and so as bidentate ligands they assist in generating large-scale architectures. Tridentate ligands such as trimeric acid are also used to bind subunits into larger structures.

More formally, an MOF is a coordination network extending, through repeating coordination entities, in one dimension, but with cross-links between two or more individual chains, loops, or spiro-links, or a coordination compound extending through repeating coordination entities in two or three-dimensions; and finally a coordination polymer is a coordination compound with repeating coordination units extending in one, two, or three-dimensions. As the chemist uses his skills and imagination to generate these 3-dimesional porous materials, they may be designed to contain pores and spaces of a particular size and shape.

In some cases, the pores are stable during elimination of the guest molecules (usually solvents) and may be used for the storage of gases such as hydrogen and

doi:10.1088/978-1-6817-4625-8ch12

carbon dioxide. Other possible applications of MOFs are in gas purification, in gas separation, in catalysis, as sensors and as super-capacitors. They are usually designed, or bespoke solids; synthesized for a particular function.

The synthesis of MOF is similar to that of zeolites, where smaller units are bound together to create a large 3-dimensional structure with particular catalytic properties. Most MOFs are made by hydrothermal or solvothermal techniques, and the crystals are grown slowly from a hot solution. The key point is that MOFs are constructed from bridging organic ligands that remain intact throughout the synthesis and act as the template, which contrasts with the synthesis of zeolites where amines are used which have to be removed later. Functionality can be built into the MOF by an appropriate choice of ligand.

Figure 12.1 gives a striking example of these new materials, MOF-5. X-ray diffraction studies reveal that this material has a cubic form (Fm3m) with $a = 25.67$ Å. The organic linker or constructing ligand is 1,4-benzenedicarboxylate (BDC) and which is seen in *ball and stick* form in the figure, and the framework formula is $Zn_4(O)(BDC)_3$. The metal cluster is a group of 4 tetrahedra (shown in blue) of $[ZnO_4]$ corner-shared by an $O^{2-}$ ion to give a tetrahedral arrangement of Zn around the central O. Remember, that in this image, the coloured spheres represent the voids within the solid framework. These are voids, or pores that are available to house or store large quantities of smaller molecules such as $H_2$ or $CO_2$.

The art of the synthetic chemist is revealed in the manner and imagination used to combine complex building blocks into these truly fantastic architectures. The chemists talk about supra-molecular building blocks being used with metal-organic polyhedra to generate structures with bespoke properties. For details of how these extraordinary structures are built, reference [2] is a long and detailed source, providing, procedures, technical details, recipes and is fully referenced; and is an open source document.

The structures of MOFs are truly impressive. But it is important to remember, that for all the aesthetic content of the images one can find all over the Internet of these weird and wonderful architectures, they are held together by the same forces that hold together the simpler structures we looked at earlier. Although the MOFs are a mixture of organic and inorganic materials, they are not held together entirely via strong ionic forces. There are strong covalent forces holding molecules together, as seen in organometallic compounds. Then there are the much weaker van der Waals interactions between molecules, and medium strength hydrogen bonding interactions in some such systems. Those parts of large molecules that are strongly bound in solids will not be executing large amplitude motions in the solid state, and so they will be relatively straightforward to localize from diffraction studies. However, those parts or moieties of these large molecules that are not held in place by strong forces (and there is a great deal of this weak bonding to be seen in MOFs) will be moving at even low temperatures and so will be more difficult to localize from diffraction studies. This will be a particular problem if one is interested in localizing all the hydrogen atoms and methyl groups that are present in the

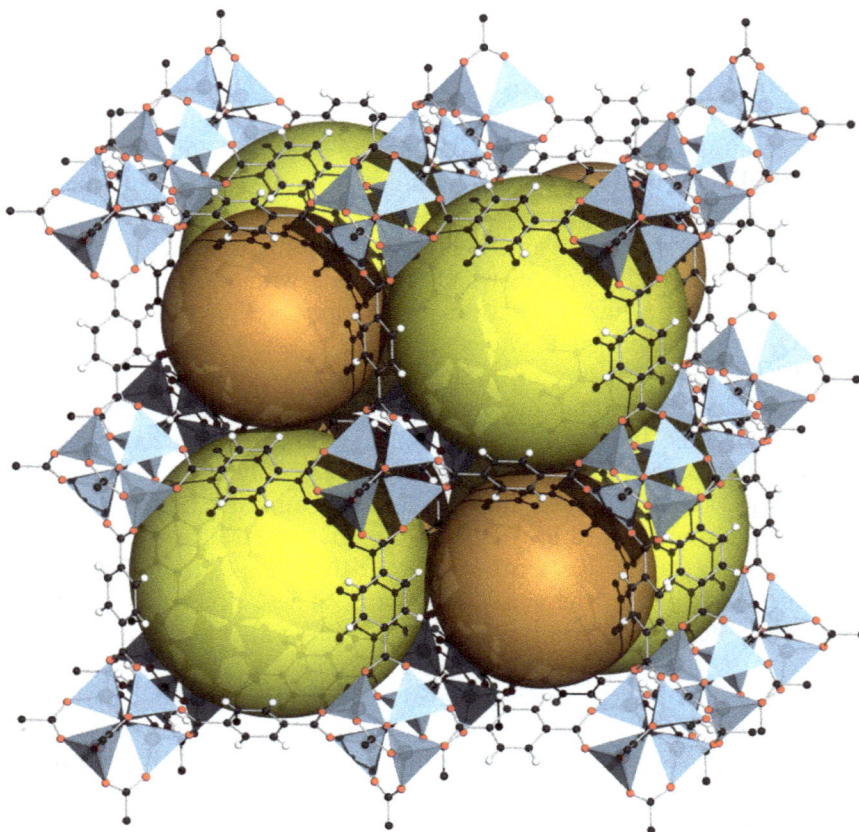

**Figure 12.1.** The structure of a particular MOF; in fact the structure of MOF-5 (a systematic nomenclature has yet to catch up with the synthetic chemists making these materials). This image is taken from https://de.wikipedia.org/wiki/MOF-5 which gives a full account of the preparation of this material. The original reference is [1]. See http://www.chemtube3d.com/solidstate/MOF-MOF5.html for an animated interactive structure, and http://www.chemtube3d.com/solidstate/MOF-home.html for a number of other animated interactive MOF structures. Point of interest: can shadows exist in molecular structures?

structure. And it is often the case that it is large amplitude motion in methyl and ethyl groups that is a prelude to a structural phase transition.

## 12.2 Deoxyribonucleic acid

'The stuff of life to knit me...' (A E Housman, from *A Shropshire Lad*, no 32)

Deoxyribonucleic acid, or DNA is the literally the *stuff of life* and one of the most famous molecules in science (and science fiction). It is a bio-molecule and a polymer, and its complex double helical structure was the subject of much controversy during the 1940s, involving some of the most well-known chemists of the day. Indeed, the controversy about the details of the structure of this molecule continued right up until the actual structure was determined by Francis Henry Compton Crick and

James Dewey Watson[1] in 1952. A discovery that changed the paradigms upon which the study of the life sciences are based, and which will still be celebrated in centuries to come when most 20th Century physicists and chemists have been forgotten.

Indeed, the unity of science has never better been illustrated than by the discovery of the structure of DNA; an event which included no little inspired guesswork, or should we say intuition on the part of Crick and Watson. A problem central to all life was solved by a combination of chemical, physical and mathematical appreciation. The arbitrariness of what may still be convenient divisions in science is reflected in the fact that the Nobel Prize awarded for the determination of the structure of DNA was the physiology and medicine prize. Knowledge of the structure of the DNA in each of us will revolutionize medicine.

The experimental technique used by Crick and Watson to unravel the structure of DNA was, as it was for nearly all the structures given in this volume, x-ray diffraction. And it is, for this author at least, always a surprise to remember that the structure of the truly huge, complex and essential DNA molecule was published in 1953 using (essentially) the same equipment and theory as was the structure of sea-salt in 1914; and in the same laboratory in Cambridge, where the director in 1952 was Sir William Lawrence Bragg.

DNA is a long polymer made up of repeated units called nucleotides. The structure of DNA is dynamic along its entire length, being capable of coiling into tight loops, and other shapes. It is found in the cells of all living things, and in all species it is composed of two helical chains bound to each other by relatively strong hydrogen bonds. Both chains are coiled round the same axis, and have the same pitch of 34 Å. The pair of chains has a radius of 10 Å. Although each individual nucleotide repeat unit consists of two smallish molecules, DNA polymers can be very large molecules containing millions to hundreds of millions of nucleotides. For instance, the DNA in the largest human chromosome (chromosome 1), consists of approximately 220 million base pairs and would be about 85 mms in length if straightened.

Figure 12.2 is a representation of the DNA molecules; however, it is only an approximate representation given the detailed structures given here of other molecules. The two, long hydrogen-bonded strands entwine to form a double helix. The nucleotide contains both a segment of the backbone of the molecule (which holds the chain together) and a nucleobase (which bonds with the other DNA strand in the helix via hydrogen bonds). A nucleobase linked to a sugar is called a nucleoside and a base linked to a sugar, and one or more phosphate groups is called a nucleotide. The sugar molecule in DNA is 2-deoxyribose, which is a pentose sugar (containing a ring of five carbon atoms and an oxygen atom; glucose is a hexose sugar). The sugars are joined together by phosphate groups that form phosphodiester bonds between the third and fifth carbon atoms of adjacent sugar rings. These asymmetric bonds mean a strand of DNA has a direction. That is, the DNA molecule has a top and a bottom, or perhaps

---

[1] Watson commented on the controversy around the structure of DNA in his book, *Some Mad Pursuit* (1988, London, Weidenfeld and Nicholson): 'No good model ever accounted for all the facts, since some data was bound to be misleading if not plain wrong'. The correct solution to the structure of DNA was a triumph of inspired imagination over a paucity of facts.

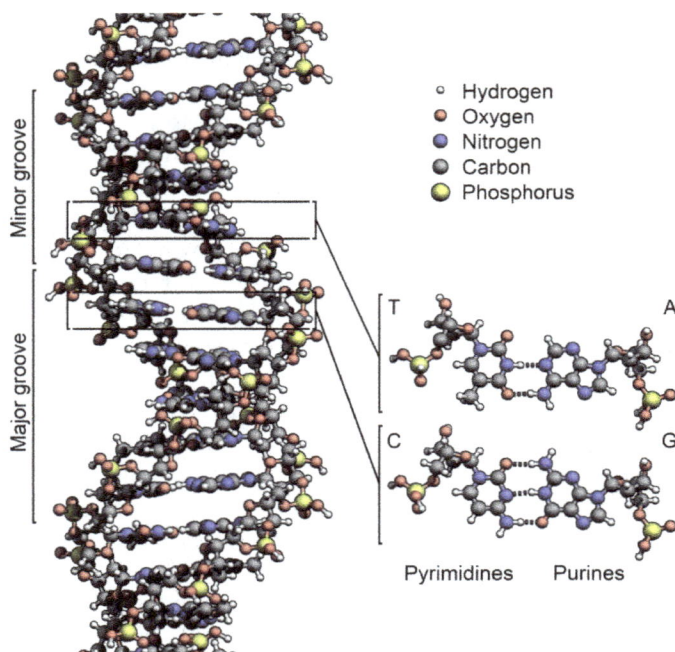

**Figure 12.2.** The structure of the DNA double helix. The atoms in the structure are colour-coded by element and the detailed structure of two base pairs is shown in the bottom right (image from Wikipedia). The architecture of this molecule is perhaps best appreciated by considering the molecule in motion https://en. wikipedia.org/wiki/DNA#/media/File:ADN_animation.gif. The nucleotide bases, the sequence of which carries our genetic details are four heterocyclic aromatic molecules; two purines and two pyrimidines. The interaction of these four molecules is shown in figure 12.3.

we should say a beginning and an end, which is why it is able to carry our genetic information from generation to generation.

The sugar and phosphate groups are located on the 'outside' of the macroscopic DNA molecules, and as these groups are strongly ionic they increase the solubility of the biopolymer in the physiological fluid of the cell by reducing the hydrophobic force. The DNA molecule is a piece of inspired design, it could not have been better engineered. But then, the DNA molecule is the result of well-over a billion years of evolution. Modern crystallographers may wish to engineer bespoke crystals, but Nature has been doing it for quite a while. In unravelling the structure of the molecules, Crick and Watson were able to explain the process of evolution by natural selection.

In a double helix, the direction of the nucleotides in one strand is opposite to their direction in the other strand; that is, the strands are antiparallel. The asymmetric ends of DNA strands are said to have a directionality of five prime (5′) and three prime (3′), with the 5′ end having a terminal phosphate group and the 3′ end a terminal hydroxyl group.

The DNA double helix is stabilized primarily by two forces: hydrogen bonds between nucleotides and base-stacking interactions among aromatic nucleobases, which have a strong basic character. In the aqueous environment of the cell, the conjugated $\pi$-bonds of the nucleotide bases align perpendicular to the axis of the DNA

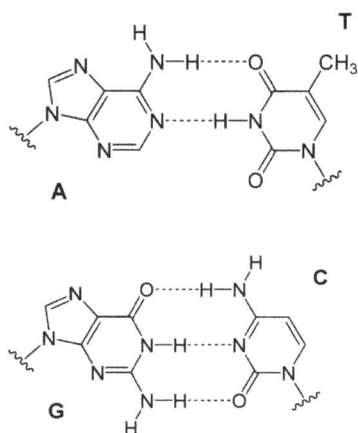

**Figure 12.3.** The structures of the bases, adenine (A), cytosine (C), guanine (G) and thymine (T), and the specific nature of their hydrogen bonding within the DNA double helix. The upper image shows a A–T base pair with two hydrogen bonds; the lower image shows a C–G base pair with three hydrogen bonds. Hydrogen bonds between the pairs are shown as dashed lines. These weak hydrogen bonds are the link between the two helical strands of the DNA.

molecule, thereby minimizing their interaction with the solvation shell and maximizing their interaction with each other. The four bases found in DNA are adenine (A), cytosine (C), guanine (G) and thymine (T), and their specific mode of interaction are detailed in figure 12.3. These four bases are attached to the sugar-phosphate to form the complete nucleotide. These four molecules are aromatic in character, and so have a large anisotropy in the charge distribution; that is, they have large electrical quadrupole moments, as seen in benzene, hexafluorobenzene and mesitylene. Indeed, the base-stacking interaction that stabilizes the large DNA molecule is identical to the forces that maintain the structure of the solid adducts benzene:hexafluorobenzene and mesitylene:hexafluorobenzene. It is found that adenine pairs only with thymine and guanine pairs only with cytosine (see figure 12.3).

This selective hydrogen-bonded pairing of the bases is termed complementary base pairing. Here, purines form hydrogen bonds to pyrimidines, with adenine bonding only to thymine via two hydrogen bonds, and cytosine bonding only to guanine via three hydrogen bonds[2]. This arrangement of two nucleotides binding

---

[2] Why should there be this combination of base-pair with two (A–T) or three (C–G) hydrogen bonds, as it is evidently the result of evolution and not of chance? Perhaps this combination of different strength binding of the two helices allows an asymmetric stiffness in the entwined hydrogen bonded strands of the double helix, which facilitates the twisting. In RNA there is no thymine, and adenine forms base-pairs with uracil (same class of molecule as thymine, except it is missing the methyl group of the thymine) but again via two hydrogen bonds. This raises another intriguing question of molecular engineering and design; why uracil and not thymine in RNA? Perhaps the loss of the methyl group (–CH₃) gives uracil the facility for generating a range of tertiary structures not needed in DNA. RNA is single stranded, but it can also coil into single-stranded loops, as in messenger-RNA. We saw earlier how the small chemical differences between benzene and toluene and ethylbenzene give these solids very different dynamics in their respective solids, but does not change other physical properties.

together across the double helix is termed a base-pair. As hydrogen bonds are not covalent, they can be broken and reformed relatively easily. The two strands of DNA in a double helix can thus be pulled apart like a zipper, either by an appropriate mechanical force (a non-bonded intermolecular force generated in the active site of an enzyme) or elevated temperature. As a result of this base-pair complementarity, all the information stored in the double-stranded sequence of a DNA helix is duplicated on each strand, which is vital in DNA replication. And thus, the mechanism of hereditary and of evolution were rationalized and made plain in 1953.

As noted above, most DNA molecules are actually two polymer strands, bound together in a helical fashion by non-covalent bonds; this double stranded structure (dsDNA) is maintained largely by the intrastrand base-stacking interactions, which are strongest for G–C stacks (with three hydrogen bonds). The two strands can separate (a process known as melting) to form two single-stranded DNA molecules (ssDNA) molecules.

The stability of the dsDNA form depends not only on the G–C content, but also on sequence (since stacking is sequence specific) and also length (longer DNA molecules are more stable). In the laboratory, the strength of this interaction can be measured by determining the temperature needed to break the hydrogen bonds between the individual strands; their melting temperature. When all the base-pairs in a DNA double helix melt, the strands separate and are present in solution, via the polarity and hydrophilic character of their ionic groups, as two entirely independent molecules. These single-stranded DNA molecules have no single common shape, but some conformations are seen to be more stable than others.

DNA is the perfect molecular machine. The product of more than a billion years of evolution. Indeed, life could not have been said to have arisen until there was DNA. So, we have a long way to go before we may emulate the perfection of the design of the DNA molecule. But chemists are busy people.

## Further reading

[1] Rosi N L, Eckert J, Eddaoudi M, Vodak D T, Kim J, O'Keefe M and Yaghi O M 2003 Hydrogen storage in microporous metal-organic frameworks *Sci.* **300** 1127– 9

[2] Guillerm V, Kim D, Eubank J F, Luebke R, Liu X, Adil K, Soo Lah M and Eddaoudi M 2014 *Chem. Soc. Rev.* **43** 6141–72 A supermolecular building approach for the design and construction of metal–organic frameworks, http://pubs.rsc.org/en/content/articlehtml/2014/cs/c4cs00135d, this is an open source document.

# Chapter 13

## Final thoughts

… all the work of the crystallographers serves only to demonstrate that there is only variety everywhere where they suppose uniformity… that in nature there is nothing absolute, nothing perfectly regular.

Georges-Louis Leclerc, Comte de Buffon (1707–1788)
in *Natural History of Minerals*, Paris 1783–88, III, 433

When beginning their study of the world around them, one of the most frequently asked questions of both chemists and physicists is: what holds solids together? Based on this question, other questions come readily to mind: why does this solid melt at a low temperature and that solid melt at a much higher temperature? What does the heat do? These questions appear simple, yet to answer them one first has to look into the solids to see how the atoms, molecules and ions are arranged. Beginning to answer such questions has been the goal of this volume.

All 27 structures given in this volume were determined by x-ray and neutron diffraction (mostly by x-ray diffraction). These are techniques that have come to revolutionize our understanding of the solid state. As mentioned earlier, it is just over a century since the Braggs determined the first crystal structure, and it has now become a routine, *black box* technology that is used by patent lawyers to argue in court (along with crystallographers as expert witnesses) about the possible infringements of a patent by a newly synthesized molecule.

Diffraction patterns such as those in figures 3.2 and 11.4(b), contain a vast amount of data. The data in figure 3.2 are profiles of scattered neutrons, and each peak represents many tens of thousands of neutrons that have been scattered into a particular angle, after having been generated in a nuclear reactor and subsequently collimated and monochromated. However, patterns such as these do not give you the structure of the solid directly. Even though they contain vast amounts of data, they are an indirect measurement [1]. This data has to be fitted to a model function, that is, an idea of what the unknown structure looks like, in order to be able to use the data to refine your initial picture of the scattering entity. Thus, you have to have a good idea of what the structure is before you can try to fit your data to the model. If the fit is poor, as determined by metrics of those data points not fitted (the residuals), then there is no fit and you have to amend your model of the structure. So, you change a bond-length, change a bond-angle, make the unit cell centrosymmetric, etc, and then try the fit again. It is an iterative process, which one hopes will converge to the true structure of the material that has scattered all those neutrons or x-ray photons. That is, a true picture of the structure that accounts for each and every scattered x-ray photon, or scattered neutron (or as many as is possible).

But given that x-ray crystallography has become a *black box* technology—indeed, the author is currently involved in a research project where my colleagues and I are using a single crystal diffractometer that does everything—you put your crystal in the appropriate place and turn it on. If everything goes to plan, within a few hours you have the 3-dimensional structure of the crystal responsible for the scattering. What the diffractometer, cum data-fitting device does not provide is an insight into why the structure is the way it is. Why the atoms of the constituent molecules are pointing this way, or that way in the solid. To unravel that mystery requires many such structural determinations, and the application of logic and the scientific method; all supported by experience. What this has demonstrated, is that even though crystallography has gone from a science in itself to being an experiment that merely tests a hypothesis, we have learnt a good deal about how crystals are engineered, how crystal architectures arise, and we are now in a position to ask a whole new set of questions.

The crystal structure gives us a picture of the balance of all the forces operating between all the atoms of all the molecules in the solid under study. To extract information about a particular long-range interaction that contributes to the attractive forces holding the crystal architecture together, we have to resort to model-based calculations.

In this text, the reader will find images and unit cell dimensions for 27 solids. There are 20 organic and organo-metallic structures, and seven structures of ionic or mixed-bonding. I began this text by discussing the electrical forces that exist between molecules, and which are the reason why the condensed phases exist. In so doing, I commented that a major contribution to the intermolecular force that causes molecules to assume a certain orientation with respect to each other in a crystal is the interaction of the first non-vanishing electrical moments of those molecules. I mentioned solid benzene and solid benzene:hexafluorobenzene (see figures 2.3 and 2.4), pointing out that the very different structures seen in these two solids may be predicted by arguments of the interaction of the molecular quadrupole moments of the molecules. This particular argument can be extended to many of the structures given here: toluene, ethane, ethylene, acetylene and nitrogen display in their solid-state what is called the herringbone structure. This is a particular arrangement of molecules arising from the interaction of molecular quadrupole moments; as defined in equation (2.5) in chapter 2 and in figure 2.2. The solid state of carbon monoxide (figure 7.2) can be rationalized by simple arguments of the interaction of the dipole moments of these molecules. This being the first non-vanishing moment in this heteronuclear diatomic.

Indeed, one could almost say that when a molecule has a certain flatness and shape anisotropy; that it is strongly 2-dimensional, then it is likely to form a herringbone structure. But then these types of molecules often have large electric quadrupole moments.

The observed structures of ionic materials are more difficult to rationalize by simple models of intermolecular electrostatics—as there are no discernible molecules. However, for the structure of the purest of ionic materials, such as sodium fluoride, these can be explained by saying that the solid is attempting to maximize the packing of spherical electrical charges so as to minimize repulsion and maximize attraction via Coulomb's Law. In solids such as calcite (figure 6.3), however, there are induction forces at work in defining the observed structure. That is, the electric fields generated by the ions polarize the anisotropically polarizable molecular ions, and gives them a large electric dipole moment in a particular direction. And it is the cumulative interaction of these induced dipole moments and the electric field of the ions that generates the observed structure.

X-ray and neutron diffraction studies are able to tell you (provided you can fit the data to a structure) where all the atoms that constitute a solid are to be found, but it is not a technique that explains how the structure works. Neither does it tell you what holds the structure together, and whether *a priori* the structure is unstable and susceptible to phase transitions; you have to measure the diffraction pattern at a number of temperatures to see this. To understand how a solid works and whether it is stable to thermal fluctuations, one must understand how energy is transmitted though the solid. In particular, how the molecules and the non-bonded intermolecular interactions are excited by the thermal background. In Nature, all is motion [2].

However, for all the ideas of what may become of the science of crystallography, its triumph is to have determined the precise equilibrium arrangements of atoms in solids. Infrared and Raman spectroscopy tells us about the dynamical structure of solids, as the infrared and laser photons are absorbed and emitted by the vibrating structures in the solid on a timescale appropriate for such actions—tens to thousands of wavenumbers, that is, $10^{10}$–$10^{13}$ s$^{-1}$, vibrational spectroscopy of solids informs us about how the solid breathes, and how the binding of the molecules varies with temperature. But x-ray diffraction generates a time-averaged, equilibrium structure, and contains a lot less dynamical information than traditional spectroscopy. However, from a combination of diffraction measurements and spectroscopy one is able to tell a great deal about the origins of crystal architecture, and of what holds crystals together. And in the case outlined in most detail here, the adduct benzene: hexafluorobenzene measurements allow us to say something definite about the nature of the lowest temperature phase transition seen in this solid, and the absence of any phase transitions below the melting points of the respective solids composed of the components of the adduct.

Whereas in pure solid benzene and in pure solid hexafluorobenzene, where quadrupole moments of like polarity interact to give a solid composed of slipped parallel and T-ordered pairs of molecules (the herringbone structure), in benzene:

hexafluorobenzene, the opposite polarity of the quadrupole–quadrupole interaction produces stacked chains of alternating, parallel benzene and hexafluorobenzene molecules. The former structure is more stable to thermal excitations than the latter, which is particularly sensitive to excitations transverse to the main chain axis. Just as bricks stacked one on top of another in parallel columns is not a sound technique for constructing buildings, it also fails to give stable extended molecular structures because of its susceptibility to vibrations perpendicular to the column axis. Solid benzene, on the other hand, having a more complex structure of overlapping and interpenetrating layers of molecules has greater stability, even with the same level of individual molecular librational excitation. It is not so susceptible to lattice distortions, and has no phase transitions below its melting point.

These investigations demonstrate that it is possible to interpret the structure and the dynamics of benzene:hexafluorobenzene and related systems, in terms of the electrostatic interactions of the constituent molecules. The ability to understand the origin of the lattice instabilities in benzene:hexafluorobenzene allows us to make predictions about other aromatic layered compounds. A survey of the literature shows that most of the organic, *donor–acceptor*, or charge-transfer, or aromatic binary adducts have at least one solid state phase transition in the same temperature range as benzene:hexafluorobenzene. Such insights into the organization and behaviour of polyatomic molecules in crystal structures has profound implications for our ability to design and produce new materials with unique mechanical, optical or electronic properties.

## References

[1] Williams J H 2016 *Quantifying measurement: The tyranny of numbers* (San Rafael, CA: Morgan and Claypool) chapters 2 and 10
[2] Williams J H 2015 *Order from Force: A natural history of the vacuum* (San Rafael, CA: Morgan and Claypool)